Moving through modernity

Moving through modernity

Space and geography in modernism

Andrew Thacker

Manchester University Press
Manchester and New York
distributed exclusively in the USA by Palgrave

Published by Manchester University Press
Oxford Road, Manchester M13 9NR, UK
and Room 400, 175 Fifth Avenue, New York, NY 10010, USA
www.manchesteruniversitypress.co.uk

Distributed exclusively in the USA by
Palgrave, 175 Fifth Avenue, New York NY 10010, USA

Distributed exclusively in Canada by
UBC Press, University of British Columbia, 2029 West Mall,
Vancouver, BC, Canada V6T 1Z2

British Library Cataloguing-in-Publication Data
A catalogue record for this book is available from the British Library

Library of Congress Cataloging-in-Publication Data
A catalog record for this book is available from the Library of Congress

ISBN 13: 978 0 7190 8120 0

First published 2003 by Manchester University Press
First paperback edition 2009

Printed by Lightning Source

Contents

To my parents, Irene and John

Figures

Acknowledgements

I wish to thank the following people for their help, advice and comments during the lengthy writing of this book: Trevor Coombs; Andy Cooper; Barbara Crowther; Derek Duncan; Maud Ellmann; Mair Evans; Peter Nicholls; Tony Pinkney; Deborah Ryan; James Ryan; Lyndsey Stonebridge; Tim Woods; Terence Hawkes, for help with the initial proposal; colleagues in English studies at the universities of Wolverhampton and Ulster; Alan Bairner and Colin Harper, for help on non-modernist subjects; Janet Campbell, Pauline Knox and Rosemary Moore for secretarial help; and Bob and Richard Thacker for technical support.

The following people provided many excellent comments (most of which I followed) on earlier drafts of chapters: Anna Cutler; Geoff Gilbert; Sally Ledger; Dick Leith; Kathleen McCracken; Scott McCracken; Joe McMinn; Sean O'Connell; David Peters Corbett; Bryony Randall; Marion Thain and Maria Wakely. My thanks go also to the library staff of a number of institutions for assistance with various queries: University of Wolverhampton; University of Ulster at Jordanstown; the British Library; Birmingham Central Library; University of Tulsa Jean Rhys Collection; London's Transport Museum; and Bet Inglis and Joy Eldridge at the Virginia Woolf Archive, University of Sussex. I have received many useful comments on papers delivered at the universities of Birmingham, Manchester Metropolitan, Salford and York, the London Modernism Seminar and various conferences, particularly the participants in the 'Modernism and the city' seminar of the Modernist Studies Association Conference 2000.

I thank the Arts and Humanities Research Board for an award of research leave during 2000–1 to enable me to complete the book, and to the Faculty of Humanities for study leave during the same period. Thanks also to Matthew Frost at MUP for his interest, enthusiasm and patience.

My biggest thanks are to Moya Lloyd, who has read and commented on enough words about modernism and space to last her a lifetime; I hope to repay her for her love, help and constant support. Daniel Thacker, my second biggest love, has my thanks for reminding me that 'exuberance is beauty' and that technology can be very, very interesting.

Part of chapter 2 appeared, in a slightly different form, as 'E. M. Forster and the motor car' in *Literature and History* 9: 2 (autumn 2000); an earlier version of chapter 3 appeared as 'Imagist travels in modernist space' in *Textual Practice* 7: 2 (summer 1993); and a section of chapter 4 was published, in a different form, as 'Toppling masonry and textual space: Nelson's pillar and spatial politics in *Ulysses*' in *Irish Studies Review* 8: 2 (2000). I am grateful to the editors for their permission to reproduce this work.

Abbreviations

HE	*Howards End*, E. M. Forster
JR	*Jacob's Room*, Virginia Woolf
MD	*Mrs Dalloway*, Virginia Woolf
O	*Orlando*, Virginia Woolf
OS	'Of other spaces', Michel Foucault
PE	*The Practice of Everyday Life*, Michel de Certeau
PS	*The Production of Space*, Henri Lefebvre
U	*Ulysses*, James Joyce
VD	*Voyage in the Dark*, Jean Rhys

Introduction: geographies of modernism

'We live in spacious times.' This bold claim seems to capture how a vocabulary of spatial and geographical terms is becoming increasingly familiar to those working in literary studies. Much work on postmodern writing, for instance, is indebted to Fredric Jameson's spatial turn ('always historicise' perhaps being replaced by a newer slogan for Jameson: always spatialise!).[1] Jameson's influential notion of a 'cognitive map' of postmodernity is, of course, a term derived from the work of the urban geographer Kevin Lynch. For Jameson also one significant difference between modernism and postmodernism lies in their relationships to the concepts of space and time:

> We have often been told, however, that we now inhabit the synchronic rather than the diachronic, and I think it is at least empirically arguable that our daily life, our psychic experience, our cultural languages, are today dominated by categories of space rather than by categories of time, as in the preceding period of high modernism.[2]

If we agree at all with this claim, then it is not surprising that we find many strands of literary and cultural studies being reoriented towards spatial questions, since it reflects only the position and the social world occupied by the contemporary critic.[3]

In postcolonial theory and criticism there is considerable attention paid to the political consequences of geographical conquest by imperialism. If imperialism and colonisation were projects intrinsically concerned with the politics of space, then it appears inevitable that we should discuss how writers produce texts that map empire, and of how resistant narratives attempt the rewriting of imposed cartographies. Edward Said, for instance, characterises his aim in *Culture and Imperialism* as 'a kind of geographical inquiry into historical experience', and claims the book demonstrates how 'the struggle over geography' crucially infuses the cultural forms of imperial power.[4]

Interdisciplinary research has also reinforced this crossover between the cultural and the geographic, with a now extensive body of work by geographers such as Derek Gregory, Doreen Massey and Steve Pile that adapts conceptual frameworks from theorists such as Foucault, Lacan, or Deleuze and Guattari. Franco Moretti's recent book on the nineteenth century novel, *An Atlas of the European Novel*, shows this critical cross-fertilisation in reverse: Moretti's stimulating book reads literary texts spatially through a series of maps depicting various locations germane to the novels. Moretti calls this methodology 'literary geography', an approach which attempts to show how 'geography shapes the narrative structure' of the novel.[5]

However, if space and geography are important theoretical orientations today, then perhaps they only recapitulate some of the central concerns of modernism. Jameson's distinction between a diachronic modernism and a synchronic postmodernism looks a little overstated once we realise that the claim 'We live in spacious times' comes not from some postmodernist or postcolonial theorist but from the modernist writer and critic Ford Madox Ford.[6] It is from his fascinating 1905 book *The Soul of London*, Ford's impressionistic attempt to 'get the atmosphere' of London.[7] Ford's book is part of that long critical tradition that analyses the urban character of modernity.[8] More interestingly, Ford's intuition concerning the spatiality of modern life occurs in a chapter devoted to means of transport into London: by motorcar, electric tram or railway.

It is that kind of connection, between space, geography and movement, in modernist writing from around 1910 to 1939 which is the central focus of *Moving Through Modernity*. Although questions of space and geography have achieved a higher theoretical profile over the last few years, relatively little work, thus far, has been done that specifically locates modernism within a renewed set of spatial or geographic contexts.[9] It is useful, therefore, to outline some broad theoretical points that orientate this book's spatial conception of modernism.

There has been an enduring tradition within critical discussions of modernism that privileges the experience and representation of temporality.[10] From Proust's meditations upon memory to the employment of the 'stream of consciousness' narrative technique by writers such as Woolf, Richardson and Joyce, or from T. S. Eliot's obsession with time and tradition in *The Waste Land* and *Four Quartets* to Pound's claim in relation to *The Cantos* that an epic is a poem which 'includes history', it seems that temporality and history are the dominant themes that plague, torment and enrich modernist writing. Wyndham Lewis's *Time and Western Man* (1927), for instance, argued that modernist writing fell foul of an obsessive interest in the 'flow' of time, to the extent of ignoring the spatial characteristics of narrative and culture.[11] This is not my argument, but Lewis's polemic indicates that we cannot easily disentangle time from space or history from geography – as some scholars of modernism have attempted to do. Indeed, the idea of 'the modern' already implies a certain temporality that distinguishes it from the non-modern.[12] As David Harvey notes, since 'modernity is about the experience of progress through modernization, writings on that theme have tended to emphasize temporality, the process of *becoming*, rather than *being* in space and place'.[13] This book argues that although temporality is clearly a significant factor in understanding the modernist project, discussions of modernism must now consider also the very profound ways in which *space, place* and *geography* occupied the modernist imagination. How do we think about the 'spacious times' of modernism using concepts of a geographical hue? It is useful to try to clarify a number of conceptual problems in any projected critical literary geography of modernism.

The first revolves around the metaphorical nature of the spaces being discussed in literary and cultural studies. To a geographer like Neil Smith the use, in certain theoretical discourses, of spatial metaphors– such as mapping, margins–centre, deterritorialisation, or location – operate at the expense of analysing the material spaces of, for example, the city. For Smith, spatial metaphors have the tendency to view actual spaces as dead or empty containers in which all objects or events can be located. He suggests that we seek to understand how metaphorical and material spaces are 'mutually implicated', and to view space not as a neutral canvas but as 'social space', produced according to social aims and objectives, and which then, in turn, shapes social life.[14] This book argues that we should understand modernist texts as creating metaphorical spaces that try to make sense of the material spaces of modernity. Chapter 1 considers some of the ways in which this approach might progress by discussing work by theorists of spatiality such as Lefebvre, Foucault and de Certeau.

The second problem concerns the representation of space in cultural texts. I argue in this book that we should think in rather more complex ways about how a text represents space, and to do this we might consider the work of the geographer Henri Lefebvre and his distinction between the *representation of space* and *representational spaces*. Lefebvre's sense of social space, discussed in chapter 1, is very broad: both internal and external, the space of the psyche, the body, the city, the house, or the room. By using Lefebvre and other cultural geographers we can analyse in more detail, for example, the nature of the specific cities encountered in modernism: Dublin in *Ulysses*; London in *Howards End*. Not only the cities in a general sense, however, but specific places within them become our concern if we inject a more developed geographical focus into the study of modernism. Now we might consider how specific streets, stations, cafés, monuments or shops are represented in modernism, and how such places offer an endorsement or contestation of official representations of space. The debates within geographical theory over the problematical distinction between space and place, considered in chapter 1, are also illuminating for how we conceive literary and cultural texts to represent space.

The third problem considers the implications of such representations for the formal properties of modernism. One of the oldest considerations of space and modernism is that of Joseph Frank in his pioneering and controversial discussion of spatial form in modernist writing, published first in the 1940s.[15] Frank's concept of spatial form is an intrinsic theoretical approach indebted to New Criticism: space is conceived as the spread of text upon paper and page, or the narrative pattern of a text read through time. In Frank there is no real discussion of *social* space: the spaces of geographical analysis, or even the spaces of literary institutions such as publishers or magazines. Frank thus ignores the kinds of space that are represented in modernist texts. But Frank's work is salutary, I think, when we consider how to discuss the representation of urban and other spaces in modernism. We need to reconnect the representational spaces in modernist texts not only to the material spaces of the city, but also to reverse the focus, and try to understand how social spaces dialogically help fashion the literary *forms* of the modernist text. This would reconnect Frank's questions about the spatial form of modernist writing to the social spaces, such as those of the city, that are often obsessively figured in modernism, tracing how social space intrudes upon the construction of the literary space of the modernist text. Literary texts represent social spaces, but social space shapes literary forms. Throughout this book I use the term *textual space* to refer to this interaction between spatial forms and social space in the literary text. Emphasis is thus given to the spatial features of literature, such as the typography and layout on the page; the space of metaphor and the shifting between different senses of space within a text; or the very shape of narrative forms, found in open-ended fictions or novels that utilise circular patterns for stories.[16] Discussion of how the formal features of literature are influenced by social or historical circumstances are always fraught with difficulty; the links in this book between space, geography and literary forms are no less tentative. Despite this, it is important, I argue, not only to discuss space and geography thematically, but to address them as questions which have a profound impact on how modernist texts are formally assembled.

To focus upon space and geography in modernism, however, does not result in a rejection of history; rather I wish to pursue an investigation into the spatial history of modernism, an account of the precise historical fashion in which particular spaces and places were conceptualised and represented. In *Postmodern Geographies* Soja argues that the reassertion of social space in cultural geography entails that we situate the spaces and places we examine within strong historical frameworks.[17] Clearly there is no sense in trying to understand how a modernist text responds to the creation or adaptation of a particular location without grasping that both social space and literary space operate in relationship to historical co-ordinates. Social space, as Soja argues, is dialectically related to history and time, and any reassertion of spatial concepts should not be a simplistic privileging of space over time. *Moving Through Modernity* is, therefore, guided by a form of critical literary geography, a methodology that draws on concepts such as Paul Carter's 'spatial history' and Kristin Ross's 'synchronic history'.[18] Such a literary geography would seek out the historical links between modernism and the production of particular material spaces in modernity. Chapter 2 examines, for example, how Forster depicts the growth of the suburbs in *Howards End*; chapter 3 interprets Imagist poetry alongside the growth of the London underground train network; while chapter 4 explores how Joyce, in *Ulysses*, interrogates the imperial history of Dublin's geography.

Thinking spatially about modernism involves recognition of the diverse ways in which 'space' might be applied to modernist texts.[19] At times it seems as if the term is so semantically vague as to be shorn of all value in critical discussion. However, this problem can be overcome by staying attentive to the precise nuances within spatial vocabularies, recalling Smith's warning about the relation of metaphorical to material spaces. We can consider, for example, psychic space, taking Freud's topographical model of the mind as one starting point for understanding how our inner life can be imagined as a set of spaces that must be related and connected in a way that is meaningful to ourselves, much as a house must have rooms and spaces that connect in some purposeful way.[20] Stream of consciousness technique in modernist fiction has, quite rightly, long been associated with philosophical theories of time and history, such as that of Bergson;[21] but it also represents a model to explore the psychic spaces of character. Narrative techniques such as interior monologue thus offer a method for moving between inner thoughts and outer reality, an approach requiring another sense of spatial terms: *inner*, *outer* and the *boundary* between these. We also need to consider how the interiority of psychic space is often profoundly informed by exterior social spaces. The chapters on Joyce, Woolf and Rhys in particular examine this interplay between interior and exterior space.

Much of modernism, then, moves away from a purely psychic perspective, introducing a whole range of other spaces. Many modernist texts – perhaps as a formal development of naturalist drama – are based in, or make great symbolic use of, rooms and domestic space.[22] There is the political longing expressed in Woolf's *A Room of One's Own*; the room of confinement in Charlotte Perkins Gilman's *The Yellow Wallpaper*; or, to take an earlier example, the door of escape slammed by Nora Helmer in Ibsen's *A Doll's House*. Moving on again, we can note the streets and buildings of the metropolis as the setting for many key modernist texts, such as the perambulations of Leopold Bloom in Dublin, or of Clarissa Dalloway in London. It is significant that city streets and domestic rooms open up the issue of the gendering of spaces in modernism, including critical discussions of the *flâneur–flâneuse*. Chapters 3, 5 and 6 consider the gendering of space in the writings of Imagist poets, and of Woolf and Rhys. A related theme explored is the gendered politics of 'the gaze' and looking in the spaces of the city, discussed in chapters 3 and 4 in particular.

In contrast to the familiar metropolitan flavours of modernism is the often occluded space of the countryside, surviving seemingly as a place of nostalgic refuge for someone like Forster in *Howards End*, discussed in chapter 2. Also of significance for Forster is *national space*, and how such spaces overlap with the geographies of imperialism and colonialism. European modernism relied, in important ways, upon the imperial spaces of Africa, the West Indies and Asia, specifically India. The works of Forster and Rhys, examined in chapters 2 and 6, overtly explore these geographical connections, showing how our critical understanding of modernism must involve the various journeys across and between 'first world' metropolitan spaces and 'third world' imperial spaces.

The arguments of critics such as Said, and many others, over the interrelationships between the great triad of modernism, the metropolis and imperialism highlight how relations between spaces are manifestly relations of power, between the occupancy and dispossession of actual geographical locations in the context of national and international politics. But such relations of power are also registered in the psychic, urban or domestic spaces described earlier. As Foucault noted in 1977: 'A whole history remains to be written of *spaces* – which would at the same time be the history of *powers* ... from the great strategies of geopolitics to the little tactics of the habitat'.[23] Geographical conflict, suggests Said, is 'complex and interesting because it is not only about soldiers and cannons but also about ideas, about forms, about images and imaginings'.[24] And these struggles take place within the literary and cultural texts of modernism, within the rooms, streets, cities and minds represented in the great modernist writing of the early twentieth century.

The switch from rooms to geopolitics, and back again, demonstrates another key characteristic of the modernist engagement with space. The multiple forms of space and geography discussed above cannot, it seems, be kept apart, even though there is, for many writers, a desperate desire to maintain borders and boundaries: rooms bleed into streets, anguished minds migrate to lands overseas. What this produces in the modernist text is that keenly felt sense of disorientation – at once both thrilling and anxious – where, for example, the psychic speculations of a person walking a city street are superimposed upon the national spaces of a great imperial power, as in Peter Walsh's juxtaposition of private thoughts, metropolitan streets and British imperialism in India in a single short passage in *Mrs Dalloway*.[25] Modernist textual space, therefore, registers these diverse social spaces: it traces the various movements between and across them; and it tries to find formal strategies to represent these disorientating, thrilling and anxious kinds of experiences. Much of this book focuses upon this polytopic quality of modernist writing.

Movement between these various spaces, then, is a key feature of modernism, and one significant way of interpreting this is via the emergence of modern means and systems of transport, such as the motorcar, the electric tram or bus, or the underground railway.[26] Analysing the significance of transport in Britain in the early years of the twentieth century enables us to understand the spaces of modernity in a more materialist fashion, as called for by Neil Smith. We can also consider how the quotidian experience of moving around the metropolis provided a key impetus to some of the experimental forms of modernist writing.[27] In this way we can develop a more nuanced account of the spatial history of modernism. It is not so much the *flâneur*, then, but more the *voyageur* that is discussed throughout this book.

The impact of the motorcar, for example, was pronounced: one early commentator proclaimed that the motorcar 'will revolutionise the world ... All our conceptions of locomotion, of transport, of speed, of danger, of safety will be changed'.[28] Artists and writers were quick to recognise the revolution in modern transport. The painter Fernand Léger noted how movement through a landscape by automobile or express train initiates a new set of sensory relations to the space perceived by the artist: 'The condensation of the modern picture, its variety, its breaking up of forms, are the result of all this. It is certain that the evolution of means of locomotion, and their speed, have something to do with the new way of seeing.'[29] By 1914 this opinion was widely accepted, with the Italian Futurists being perhaps the most notable of the modernist groups to embrace the euphoria accompanying the automobile. Wyndham Lewis sniffily but perhaps accurately dismissed the Futurists as 'Automobilists', while proclaiming that his Vorticist group 'blessed' another transportative technology, the 'restless machinery' of English shipping.[30] As well as the general social revolution of technologies like the motorcar, the impact of transport upon literature was also noted. I. A. Richards complained in *Practical Criticism* (1929): 'No one at all sensitive to rhythm, for example, will doubt that the new pervasive, almost ceaseless, mutter and roar of modern transport, replacing the rhythm of the footstep or of horses' hoofs, is capable of interfering in many ways with our reading of verse.'[31] Richards has a footnote to this comment citing T. S. Eliot in support of his claim that the 'internal combustion engine may already have altered our perceptions of rhythms'. Eliot had himself already linked modernity and modernism to transport when, in 1921, he likened the music of Stravinsky's *The Rite of Spring* to the 'the scream of the motor-horn, the rattle of machinery, the grind of wheels, the beating of iron and steel, the roar of the underground railway, and the other barbaric noises of modern life'.[32]

It is this attention to the machinery of modernity, barbaric or euphoric, that thus underpins my choice of title: *Moving Through Modernity*. For it tries to capture the sense that modernist writing can be located only within the movements between and across multiple sorts of space. This is a movement through new material spaces and by means of the new machines of modernity, and which grounds a more abstract sense of flux and change that many modernist writers attempted to articulate in their texts: Forster, for example, has one of his characters bemoan 'this continual flux of London'.[33] This book comprehends that lament by relating the notion of 'flux' to material motion through specific spaces and geographies. In the new *topoi* of the early twentieth century, transportation emphasised a sense of movement that came to be a crucial figure for the experience of modernity itself.[34]

This book, then, considers the complex relationships between spaces, modernity and modernism. Modernity, as Foucault once suggested, is perhaps best understood, not as an epoch or period, but as a certain *attitude* towards the historical present.[35] A crucial component of how modernists regarded the present was their attitude to space and geography. This book links together an understanding of space as both material and metaphor: the space of the literary text, its form as narrative or poem, and how these textual spaces reveal a multifaceted range of attitudes to the spaces of modernity. Beginning with a discussion of methodological questions concerning space and place, the book then takes a roughly chronological focus upon Forster, Imagism, Joyce, Woolf and Rhys. Again for reasons of coherence and focus, I have tended to concentrate upon a limited range of texts by each author, discussing Forster's *Howards End*, for example, but little else by this writer. I have chosen these British and Irish writers because of the historical range of their writing across the modernist period, for the variety of styles of writing they employ, and for the multiple geographies of modernism they explore. This book, therefore, aims to demonstrate both the validity of a literary geographical approach to modernism, and a fuller picture of the 'spacious times' that modernism inhabited.

Notes

1 Fredric Jameson, *The Political Unconscious: Narrative as a Socially Symbolic Act* (London, Methuen, 1981), p. 9.
2 Fredric Jameson, *Postmodernism, or the Cultural Logic of Late Capitalism* (London, Verso, 1991), p. 16. 'Cognitive mapping' occurs in Kevin Lynch, *The Image of the City* (Cambridge, MA, MIT Press, 1960).
3 Recent books on contemporary literature and 'space' include Brian Jarvis, *Postmodern Cartographies: The Geographical Imagination in Contemporary American Culture* (London, Pluto, 1998); Maria Balshaw and Liam Kennedy, eds, *Urban Space and Representation* (London, Pluto, 2000); and Glenda Norquay and Gerry Smyth, eds, *Space and Place: The Geographies of Literature* (Liverpool, Liverpool John Moores University Press, 1997). For a discussion of earlier literature see *John Gillies, Shakespeare and the Geography of Difference* (Cambridge, Cambridge University Press, 1994).
4 Edward Said, *Culture and Imperialism* (London, Vintage, 1994), p. 6.
5 Franco Moretti, *Atlas of the European Novel 1800–1900* (London, Verso, 1998), p. 8. For a brief account of the prospects for a new 'literary geography' such as Moretti's, see John Kerrigan, 'The country of the mind', *Times Literary Supplement* (11 September 1998), 3–4.
6 Ford Madox Ford, *The Soul of London*, ed. Alan G. Hill, (London, Everyman, 1995), p. 41.
7 Ford, *The Soul of London*, p. 4.
8 For a recent discussion of this issue see Peter Brooker, *Modernity and Metropolis: Writing, Film and Urban Formations* (Basingstoke, Palgrave, 2002).
9 Two examples of such work are Chris GoGwilt, *The Invention of the West: Joseph Conrad and the Double-Mapping of Europe and Empire* (Stanford, CA, Stanford University Press, 1995) and Elisabeth Bronfen, *Dorothy Richardson's Art of Memory*, trans. Victoria Appelbe (Manchester, Manchester University

Press, 1999). There was also a successful panel on modernism and geography, organised by Jon Hegglund, at the Modernist Studies Association Conference in Houston, 2001.

10 For an overview of how modernism has been theorised see Astradur Eysteinsson, *The Concept of Modernism* (Ithaca, NY, and London, Cornell University Press, 1990).

11 Wyndham Lewis, *Time and Western Man* (Boston, MA, Beacon Press, 1957).

12 For discussion of the problematical idea of 'the modern' in relation to temporality see Paul De Man, 'Literary history and literary modernity' in his *Blindness and Insight: Essays in the Rhetoric of Contemporary Criticism* (London, Methuen, 1983) and Peter Nicholls, *Modernisms: A Literary Guide* (Basingstoke, Macmillan, 1995), ch. 8.

13 David Harvey, *The Condition of Postmodernity* (Oxford, Blackwell, 1990), p. 205.

14 Neil Smith, 'Homeless/global: scaling places', in Jon Bird, Barry Curtis, Tim Putnam, George Robertson and Lisa Tickner, eds, *Mapping the Futures: Local Cultures, Global Change* (London, Routledge, 1993), pp. 98–9. For a similar criticism of metaphors of travel see Janet Wolff, 'On the road again: metaphors of travel in cultural criticism' in *Resident Alien: Feminist Cultural Criticism* (Cambridge, Polity Press, 1995).

15 Joseph Frank, 'Spatial form in literature' (1945), reprinted in *The Widening Gyre: Crisis and Mastery in Modern Literature* (Bloomington and London, Indiana University Press, 1968). See also the later debate on the formalist and historical implications of the concept: Frank, 'Spatial form: an answer to critics', *Critical Inquiry* 4 (1977–78), 231–52; Frank Kermode, 'A reply to Joseph Frank', *Critical Inquiry* 4 (1977–78), 579–88; Eric S. Rabkin, 'Spatial form and plot', *Critical Inquiry* 4 (1977–78), 253–70; William Holtz, 'Spatial form in modern literature: a reconsideration', *Critical Inquiry* 4 (1977–78), 271–83; Joseph Frank, 'Spatial form: some further reflections' *Critical Inquiry* 5 (1978–79), 275–90; and W. J. T. Mitchell, 'Spatial form in literature: towards a general theory', *Critical Inquiry* 6 (1979–80), 539–67.

16 One interesting approach to modernist space would be to consider the location of some of the institutions of literature: where modernist works were published, for example, and the role of little magazines in its dissemination. For such an approach, but without an explicitly spatial focus, see Lawrence Rainey, *Institutions of Modernism: Literary Elites and Public Culture* (New Haven, CT, Yale University Press, 1998).

17 Edward Soja, *Postmodern Geographies: The Reassertion of Space in Critical Social Theory* (London and New York, Verso, 1989), pp. 23–4.

18 See Paul Carter, *The Road to Botany Bay: An Essay in Spatial History* (London, Faber, 1987); Kristin Ross, *The Emergence of Social Space: Rimbaud and the Paris Commune* (London, Macmillan, 1988).

19 Two forms of space that I have not had room to discuss here concern contemporary scientific discourses, such as the theory of relativity, and that of travel writing in the modernist period. For a consideration of the impact of scientific theories see Stephen Kern, *The Culture of Time and Space 1880–1918* (Cambridge, MA, Harvard University Press, 1983), ch. 6, and Randall Stevenson, *Modernist Fiction: An Introduction* (Hemel Hempstead, Harvester Wheatsheaf, 1992). On travel writing in this period see Paul Fussell, *Abroad: British Literary Travelling between the Wars* (Oxford, Oxford University Press, 1980) and Caren Kaplan, *Questions of Travel: Postmodern Discourses of Displacement* (Durham, NC, and London, Duke University Press, 1996), pp. 27–64.

20 Freud considered the unconscious from what he called a 'topographical point of view'; see Sigmund Freud, 'The unconscious' in *On Metapsychology: The Theory of Psychoanalysis*, Penguin Freud Library (London, Penguin, 1986), vol. 11.

21 For the impact of Bergson see Sanford Schwartz, *Matrix of Modernism: Pound, Eliot and Early Twentieth-Century Thought* (Princeton, NJ, Princeton University Press, 1985).

22 For the significance of rooms in naturalist drama and the link with modernism see Raymond Williams, 'Theatre as a political forum' in *The Politics of Modernism: Against the New Conformists* (London, Verso, 1989).

23 Michel Foucault, 'The eye of power' in *Power/Knowledge: Selected Interviews and Other Writings 1972–1977* (London, Harvester Wheatsheaf, 1980), p. 149.

24 Said, *Culture and Imperialism*, p. 6.

25 Viginia Woolf, *Mrs Dalloway* (London, Granada, 1976), p. 45. For an analysis of this passage see ch. 5, this book.

26 I have been unable to discuss here the full panoply of modes of transport in modernism: vehicles that might play a salient role include bicycles, aircraft, tanks, ships and motorbikes.

27 Ford, for example, in *The Soul of London*, concludes his chapter 'Roads into London' by drawing a link between the 'pathos and dissatisfaction' of gazing out of a train window at incidents from daily life that one never sees completed, and the desire for stories to have an ending (p. 43).

28 Joseph Pennell, 'Motors and cycles: the transition stage', *Contemporary Review* (1 February 1902), 185.

29 Fernand Léger, 'Contemporary achievements in painting' (1914), in Edward F. Fry, ed., *Cubism* (London, Thames & Hudson, 1966), p. 135.

30 Wyndham Lewis, 'The melodrama of modernity', *BLAST* 1 (1914), ed. Bradford Morrow, reprinted (Santa Rosa, CA, Black Sparrow Press, 1989), 143; Lewis, 'Manifesto', *BLAST*, 22–3.

31 I. A. Richards, *Practical Criticism: A Study of Literary Judgement* (London, Routledge, 1964), p. 318.

32 Eliot, quoted in Lyndall Gordon, *Eliot's Early Years* (Oxford, Oxford University Press, 1977), p. 108. For a general overview of artistic representations of the

motorcar in the early twentieth century see Sean O'Connell, *The Car and British Society: Class, Gender and Motoring, 1896–1939* (Manchester and New York, Manchester University Press, 1998), ch. 6.

33 E. M. Forster, *Howards End*, ed. Oliver Stallybrass (Harmondsworth, Penguin, 1983), p. 184.

34 This is not to imply that transport did not have a considerable cultural impact on writers and artists in earlier centuries: the case of Charles Dickens and the railways indicates that it did. The focus of this book is, however, on the modernist period and the specific nature of the relationship between modernist writing and transport.

35 Michel Foucault, 'What is Enlightenment?' in *The Foucault Reader*, ed. Paul Rabinow (Harmondsworth, Penguin, 1984), p. 39.

1 Theorising space and place in modernism

'This complex historical geography of modernism (a tale yet to be fully written and explained)'[1] is David Harvey's provocative description of Bradbury and McFarlane's *Modernism 1890–1930*, the hugely influential account of the emergence of modernism in different cities and countries.[2] In a modest fashion I take up that challenge by trying to develop a form of critical literary geography adequate to the complexities Harvey mentions. This chapter critically considers a number of important theories of space and place as found in Heidegger, Bachelard, Lefebvre, Foucault, de Certeau and Harvey. In discussing these writers my aim is to elaborate a critical terminology of space and place with which to begin a geographical approach to modernism. This chapter, therefore, is mainly expository, outlining a set of arguments and concepts taken from geographers and theorists that will be employed throughout this book. In addition to the broad distinction between space and place, several other spatial concepts will be considered, including Lefebvre's theory of 'social space', Foucault's conception of heterotopias, and de Certeau's distinction between the tour and the map. Throughout this chapter these ideas are related, schematically, to a number of modernist texts so that the reader can grasp how these spatial and geographical notions will, in later chapters, be more fully applied.

Modernist writing, this book argues, is about living and experiencing 'new times', not in the abstracted location of literary history, but in specific spatial histories: rooms, cities, buildings, countries and landscapes. This understanding of the multiple material geographies of modernism is guided by one of the key sets of terms found in recent geographical and spatial theory: the opposition between space and place.[3] To a number of geographical theorists *space* indicates a sense of movement, of history, of becoming, while *place* is often thought to imply a static sense of location, of being, or of dwelling. Much modernist writing oscillates between these twin spatial visions, often in ways that complicate any sharp and easy division between a conservative sense of place and a revolutionary sense of space. In considering these different attitudes to modernity – space or place, movement or location – we can start to elucidate more fully Harvey's 'complex historical geography of modernism'.

Heidegger and Bachelard

Perhaps the earliest philosophical distinction between space and place is articulated by Heidegger in 'Building, dwelling, thinking' (1951). This is a classic statement of the space–place division and is drawn upon by many other thinkers, albeit with quite distinct emphases. Heidegger discusses the relation between the physical nature of a building and the human experience of dwelling within it. *To dwell* refers to a 'manner in which we humans are on the earth'; thus '[to] be human ... means to dwell'.[4] Dwelling is thus a particular characteristic of being human, a basic ontic state, whereby one's very being is located in a particular place. Heidegger is critical of any abstract conception of space, such as is found in mathematics or physics. Spaces such as buildings receive their being from locations and the experience of those dwelling in them: 'the spaces through which we go daily are provided for by locations'.[5] We should not think of space as something external to human beings, since space is intrinsically linked to the dwelling experience of humans: 'Man's relation to locations, and through locations to spaces, inheres in his dwelling. The relationship between man and space is none other than dwelling.'[6]

Heidegger's argument is that the true nature of space – whether of the external environment, or of the body – is conditioned by this existential experience of dwelling: 'To say that mortals *are* is to say that *in dwelling* they persist through spaces by virtue of their stay among things and locations.'[7] To dwell means to preserve, indicating a certain conservatism of tradition that infuses the notion of dwelling in a place. Being itself is associated strongly with a sense of place. Heidegger thus prioritises the nature of place over space, according to critics, in his insistence that a kind of static dwelling in a particular location is the fundamental core of the relation between human beings and space. This results in place being viewed in an unhistorical and asocial fashion. For critics, Heideggerian dwelling seems to consist of a staying put, a rejection of spaces as products of history and social interaction: 'Dwelling ... is the basic character of Being in keeping with which mortals exist'.[8] It also views place as a fixed and bounded object, rather than as a site formed by specific social processes and amenable to future changes.[9] Many literary texts that eulogise places as locations of significance for the individual writer seem indebted, consciously or unconsciously, to Heidegger's conception of place.[10]

A slightly different account of space and place, but one influenced by Heidegger's work, is Gaston Bachelard's *The Poetics of Space* (1958). This quirky yet stimulating blend of phenomenology and psychoanalysis aims to uncover the primary importance of inhabiting a particular place. Unlike Heidegger's abstract sense of 'dwelling' and 'being', Bachelard's is a 'concrete metaphysics' that situates the first and primary sense of place in an actual location, that of the house.[11] Bachelard wishes to produce 'a phenomenological study of the intimate values of inside life',[12] eschewing Heidegger's imprecise notion of dwelling for the specific nooks and crannies of the house, particularly the house in which the child first comes to consciousness. Bachelard's book thus has chapters upon various aspects of the house, such as the cellar, the attic, chests and wardrobes, and the experiential sense of corners in rooms or the vertical design of a house. Rather than mere descriptions of such places, Bachelard wants to capture 'the primary function of inhabiting' that is located in the house; as he suggests, 'all really inhabited space bears the essence of the notion of home'.[13] Bachelard's term for this form of study is *topoanalysis*, and its focus is upon spaces that have been turned into places of pleasurable belonging, the transformation of a house into a home, where the rooms and corridors of the house articulate 'the topography of our intimate being'. [14] Space, writes Bachelard, 'that has been seized upon by the imagination cannot remain indifferent space subject to the measures and estimates of the surveyor'.[15] As Edward Casey comments on Bachelard's conception of the house, it is a kind of 'place-world, a world of places' where the exploration is not so much geometrical or architectural as imaginative or poetic.[16] Bachelard's examples are nearly all drawn from poetry, since this imaginative realm is closest, he argues, to that of dreams and the unconscious, where our most intimate memories of inhabiting early childhood places are stored, or 'housed'.

Bachelard, as Casey notes, provides a richer and more tangible account of the sense of place than does Heidegger, while still privileging place over space.[17] Bachelard's work is suggestive, however, in its demand that space be read as a text: we are said to 'write a room', 'read a room', or 'read a house'.[18] The book is interesting also for Bachelard's sense of body-space, his seeing the house as a kind of body. Our primary experiences of home inscribe themselves in our somatic lives, leaving memory traces of the rooms and spaces which we have inhabited in the 'passionate liaison of our bodies ... with an unforgettable house'.[19] One example, discussed in the next chapter, is the intimate intertwining of house and body in Forster's *Howards End*.

However, *The Poetics of Space* seems restricted by what is its most engaging and innovative feature, its topophilia. In its overwhelming focus upon 'quite simple images of *felicitous space*'[20] we find a conception of place as wholly benign, one which is unable to imagine conflict within the realms of intimate space. The home, which for Bachelard is a place of intimacy and warmth, can quite easily contain many dramas of conflict and unease. In Charlotte Perkins Gilman's *The Yellow Wallpaper*, for example, the bedroom is not a setting for serene intimacy, but a place of confinement where a gendered set of power relations is enacted between the imprisoned female and her doctor husband.[21]

Another limitation in Bachelard's account is his concentration on interior places. Any sense of exterior space, whether of streets, cities or nations, is left unexplored. Arguably, the sense of intimacy and inhabitation he describes as characterising the house could also be applied to exterior locations. Much of Rhys's fiction of belonging involves a set of national locations in addition to the interiors of rooms. Bachelard also does not explore the complex relations between the intimate spaces of rooms or houses and where they are found, in streets or landscapes. It is as if he can feel a tender sense of place only when imagining the interior of a house; the passionate attachment to a piece of land, a village, or a city street is never fully discussed by Bachelard.[22] Equally, Bachelard does not address questions such as how the architectural design of a house might influence one's topographic attachments, or how the social and political history of architectural forms might alter one's intimate inhabitation of a place. As we shall see, for instance when discussing *Howards End* or *Ulysses*, a sense of intimate attachment to place can be properly understood only in relation to the historical constitution of houses, buildings and spaces.

This wider understanding of how space and place interact is found in Lefebvre's *The Production of Space*, a vital theoretical text for recent cultural geography. Lefebvre's conception of 'social space' is designed to introduce questions of society, history and politics into thinking about space, and involves both a rejection of Heidegger's ontological valuation of place as a site of dwelling and an extension of Bachelard's notion of topoanalysis.

Lefebvre and social space

Henri Lefebvre's 1974 *The Production of Space* (hereafter *PS*) has been very influential in introducing a number of key concepts into spatial and geographical theory, perhaps the most significant being that of 'social space'.[23] Lefebvre argues that, for many years, 'the word "space" had a strictly geometrical meaning: the idea it evoked was simply that of an empty area' (*PS*, p. 1). This abstract view of space originated with Descartes and was influential throughout the humanities and social sciences, as well as in mathematics and the physical sciences. Lefebvre aims to reverse this widespread view by insisting that space is not a vacuum merely *containing* other objects and practices: 'space is never empty: it always embodies a meaning' (*PS*, p. 154). Since space is always produced by social practices it can always be deciphered for specific social meanings: '(Social) space is a (social) product' (*PS*, p. 26). Here Lefebvre draws on Marx's view of the role of relations of production in society, applying Marx's notion to the way that landscapes are altered by human productive practices. This dialectical view of space and society argues that every society produces its own distinctive form of space, from the ancient *polis* of the Greek world to the city-state of the Italian Renaissance, or the high-tech postmodern cities of the present.

Forms of spatial organisation, however, also play a dominant role in shaping societies, determining the realms of mental space and physical space. Most spaces, writes Lefebvre, are 'at once a precondition and a result of social superstructures' (*PS*, p. 85), and 'any space implies, contains and dissimulates social relationships' (pp. 82–3). Once we leave natural spaces such as uncultivated land, forests or heaths we enter a world where space encounters the social relations of production. A national forest park, for example, though it might appear to be an inherently natural space, assumes certain social characteristics once it is administered by the state or redesigned for visitors or tourists. It becomes a social space because of its relationship to factors such as the economic development of a particular region or a national strategy for tourism. If part of the forest is given over to the timber industry, then this again produces a very different form of social space.

For Lefebvre social space is an overtly political concept:

> Space has been shaped and moulded from historical and natural elements, but this had been a political process. Space is political and ideological. It is a product literally filled with ideologies. There is an ideology of space. Why? Because space, which seems homogenous, which seems to be completely objective in its pure form ... is a social product. The production of space can be likened to the production of any given particular type of merchandise.[24]

Lefebvre's work has thus influenced the development of a 'socio-spatial' dialectic, as Soja terms it:[25] society shapes spaces according to its needs, but, equally, space plays a formative role in the construction of social life.

Lefebvre's conception of social space, as he acknowledges, is a very broad one, with a frustrating tendency towards vagueness. However, one clear and significant feature is that social space is inherently composite, mingling heterogeneous spaces together in one physical location. As the forest park example shows, any single piece of land can be analysed into many different social spaces with quite distinct associated meanings: a place for tourists to walk; a place for timber production; an area for economic development; or a signifier of a particular regional or cultural identity (the highlands of Scotland, the bog lands of the west of Ireland). Unlike Heidegger's conception of an attachment to discrete places, Lefebvre's social spaces always intertwine with one another:

> *Social spaces interpenetrate one another and/or superimpose themselves upon one another.* They are not *things*, which have mutually limiting boundaries ... Visible boundaries, such as walls or enclosures in general, give rise for their part to an appearance of separation between spaces where in fact what exists is an ambiguous continuity. The space of a room, bedroom, house or garden may be cut off in a sense from social space by barriers and walls, by all the signs of private property, yet still remain fundamentally part of that space. (*PS*, p. 87)

This multifaceted character of space makes it particularly suitable for thinking about in modernism, given the multiple senses of literary space – from small rooms to images of the city – outlined in the Introduction. The 'ambiguous continuity' of different social spaces is a dexterous concept for the literary geographer, being particularly apt for understanding how modernism combines very distinct spaces in quite startling ways, such as Rhys's perpetual switching between England and Dominica in *Voyage in the Dark*.

Lefebvre's description of social space recalls something of Bakhtin's sense of the heteroglossic nature of language, where each word partakes of an almost infinite number of social discourses that come into contact with one another and help shape any specific meaning amid the constant flux of language.[26] Thus Lefebvre writes that a local space is not eradicated by a larger regional, national or global space, but enters into a complex set of relationships with these other spaces:

> The national and regional levels take in innumerable 'places': national space embraces the regions; and world space does not merely subsume national spaces, but even ... precipitates the formation of new national spaces through a remarkable process of fission. All these

> spaces, meanwhile, are traversed by myriad currents. The hypercomplexity of social space should by now be apparent, embracing as it does individual entities and peculiarities, relatively fixed points, movements, and flows and waves – some interpenetrating, others in conflict, and so on. (*PS*, p. 88)

A house in a street, for example, may appear to be solid and immobile as a space. Analysed as social space, however, both the house and the street are 'permeated from every direction by streams of energy which run in and out of [them]' (*PS*, p. 93), such as gas, electricity, radio and television signals. The picture of a static space is replaced by 'an image of a complex of mobilities, a nexus of in and out conduits' (p. 93). Each of these conduits opens this space out into wider social, political and economic questions and concerns. For Lefebvre, then, place is not to be sharply distinguished from space. Rather, a writer's conception of some particular place should be understood in relation to the wider historical and social meanings of that site. Place is ultimately, for Lefebvre, only one form, though with its own ideology and politics, of the many existing discourses of social space. For Lefebvre, the Heideggerian view of place as a universal dwelling for human *being* embodies only a particular, and historically specific, view of social space and social relationships.[27]

Polytopic texts like Joyce's *Ulysses* or Eliot's *The Waste Land* clearly display 'hypercomplex' social spaces, with conflicting relations between different spaces being a key element in the formation of their textual spaces. A poem like *The Waste Land* that represents morbid consciousnesses, soiled bodies and dingy rooms in suburban streets, along with references to beaches, mountains, tube stations and other countries, is thus particularly amenable to a Lefebvrean reading. Analysis should attend not only to the particular significance of each discrete space, but to their interconnection and the quarrels between them. This principle of the 'interpenetration and superimposition of social spaces' entails that any 'fragment of space' under analysis will reveal not one but many social relationships. Bachelard's topoanalysis of intimate spaces can, therefore, be usefully augmented by Lefebvre's notion of the hypercomplexity of social space.

In *The Production of Space* Lefebvre introduces three aspects of social space that are of particular utility for a literary geography of modernism:

- *Spatial practices* refer to the multiple activities that form spaces in each society, embracing features such as production and reproduction; it also refers to the spatial actions of each individual in a society – everyday journeys to work, to home, to sites of leisure or consumption, the roads and the transport practices that dominate material life. Lefebvre refers to these as *experienced* spaces, to indicate how an individual practically relates to the outside world. Broadly, it refers to what people do in spaces.

- *Representations of space* are linked to official relations of production and order; this is space as *perceived* by planners, architects and governments, and is the dominant space in any society. Though abstract in nature, drawn on plans, maps and diagrams, representations of space are social and political in practice and considerably alter the production of space through, for example, the construction of a monument, a national museum or schemes for road-widening in a city. Representations of space modify the spatial texture of a city or landscape according to certain ideologies, and are linked to codes and signs – for example, those used on a proposed redevelopment plan of a city site.
- *Representational spaces* embody space as *imagined* by inhabitants, and is often linked to artists and writers, and to 'the clandestine or underground side of social life' (*PS*, p. 33). It refers to the 'dominated … space which the imagination seeks to change and appropriate' (p. 39). Representational spaces do not directly alter the construction, for example, of a city, but rather result in symbolic and artistic productions. Representational space refuses the rational order and cool logic of representations of space; instead representational space 'is alive: it speaks. It has an affective kernel or centre: Ego, bed, bedroom, dwelling, house; or: square, church, graveyard. It embraces the loci of passion, of action and of lived situations, and thus immediately implies time' (*PS*, p. 42).[28]

These three categories for understanding social space are, as we might expect, interconnected. Lefebvre focuses upon 'the fragmented and uncertain connection between elaborated representations of space … and representational spaces' as well as the spatial practices of people living within these two forms of representation (*PS*, p. 230). Frustratingly, Lefebvre does not offer many elaborated examples of how the experienced–perceived–imagined triad interacts.[29] But there are a number of ways in which Lefebvre's theories can be utilised in the study of modernism. The focus on the symbolic qualities of representational spaces can be directly applied to modernist writers who attempted to capture the new spaces of modernity in their works. Modernism was engaged in a diverse set of responses to the official representations of space in modernity, found in new forms of urban life such as the suburb and transport systems, or in relations between the imperial capital and the colony. Some writers celebrated the rational forms and structured logic of these representations of space, while others found much to criticise or challenge in them, as we will see. Lefebvre's focus upon the hypercomplexity of social space is a good reminder, however, of the myriad responses to space that we find in modernist writing: to homogenise them into a monolithic representational space of modernism is to ignore the contradictory ways in which space and place were conceived and imagined by different writers.

Lefebvre's work also stresses that the representational spaces found in literary texts are to be connected to material spaces and places, and to the representations of space embodied in them. To study, for instance, the London of *Mrs Dalloway*, the Dublin of *Ulysses* or the Dominica of *Voyage in the Dark* is to consider the material spaces these texts discuss, and to disclose how the representational spaces of these texts reflect, contest or endorse the geographical shaping of these *topoi* by various ideological representations of space. Very often we might consider the ways in which writers appropriate spaces dominated by official meanings, producing representational spaces with quite different meanings, such as Joyce's revisioning of Dublin's status as an imperial outpost of the British Empire.[30] Lefebvre's work thus emphasises how a spatialised reading of modernism must always consider relations of power and geopolitics.[31]

Scattered throughout *The Production of Space* are hints about the nature of space in modernism and the dominant representation of space in modernity. Modernity, for Lefebvre, is described as *abstract space*, a representation of space from the eighteenth century onwards that wishes to homogenise social space. Abstract space is formal and quantitative, and functions by regarding space as an object filled with materials such as glass and stone, concrete and steel (*PS*, p. 49). Lefebvre discusses the role of the Bauhaus architectural group in codifying the central tenets of abstract space in the 1920s, forming a direct link between the modernist avant-garde and the space of modernity (*PS*, pp. 124–7). Abstract space is characterised by three more features: its reliance upon geometry; a stress upon a 'logic of visualisation', where the visual gaze predominates over any other sensual feature of the human body; and a desire to fill the empty homogenised space it desires with what Lefebvre terms a 'phallic verticality' (*PS*, p. 287), an expression of masculine violence associated with bureaucracy and the state. Glass skyscrapers and towers might be said to epitomise Lefebvre's image of this abstract space.

Lefebvre himself was convinced that art and literature reveal a profound understanding of these new social spaces of modernity. Of Picasso's great innovations around 1907, with the painting of *Les Demoiselles d'Avignon* for example, he writes: 'Picasso's space *heralded* the space of modernity' (*PS*, p. 302), an abstract space dominated by visuality and phallic violence. Picasso's Cubism is thus linked to the dominant representation of space found in the abstract space of modernity: social space is intertwined with the aesthetics of spatial form. Later chapters considering Imagist poetry around 1909–10, and the troubled encounter with modernity found in Forster's *Howards End*, published in 1910, will be guided by Lefebvre's linkage of social space to aesthetic space, and by his observation that in Europe 'around 1910 a certain space was shattered' (*PS*, p. 25), the representation of space found in Euclidean and perspectival thought from the European Renaissance onwards. Within the destruction of this space we witness the genesis of the new textual spaces of modernism.

Lefebvre's suggestion that the spatial forms of modernist culture must be connected to alterations in material space is an important one. One theorist who extends this view in a particularly interesting fashion is Michel Foucault. Foucault's unique sense of space, which combines both metaphorical and material dimensions, is most noticeable in his conception of *heterotopia*.

Foucault and heterotopia

Foucault wrote specifically on spatiality and geography only minimally, but the influence of such writings has, perhaps surprisingly, been widespread.[32] We can distinguish three broad ways in which these ideas have been significant. First, there are his scattered comments on questions of geography and space – particularly in relation to topics such as the spatial organisation of power in prisons and the architecture of schools – or on the body. This we might characterise as an interest in a Lefebvrean social space, or what Foucault calls 'external spaces' ('Of other spaces' (hereafter OS), p. 23).[33] Second, Foucault employs a variety of spatial metaphors in his work, terms such as 'site', 'field', 'domain', 'grid', 'strata' or 'epistemological space'. Edward Said, for example, refers to Foucault's 'strategic and geographical sense' of method, a more spatial mode than the temporal sense of Hegelian or deconstructive forms of theory; while John Marks describes Foucault's 'broadly spatial approach to thought' itself.[34] A third aspect of Foucault's spatial theory is his concept of heterotopia, which is a provocative way of combining material and metaphorical senses of space.

For Foucault, deciphering discourses 'through the use of spatial, strategic metaphors enables one to grasp precisely the point at which discourses are transformed in, through and on the basis of relations of power'.[35] In this way Foucault couples his vocabulary of spatial metaphors to an analysis of material spaces, using power as a tool to unpick the 'social apparatus' (*dispositif*) of, for example, the prison or sexuality.[36] In this sense Foucault's work traces how spatial metaphors and material spaces interact on the basis of relations of power. For example, the panopticon in *Discipline and Punish* is simultaneously a material presence and a trope for the gaze of disciplinary power in modern societies.[37]

Power is, then, the key for shifting Foucault's spatial imagination away from the two-dimensional plane of structuralism and into a fully social and historical three-dimensional space.[38] In a 1977 interview Foucault discussed the nature of domestic architecture and how spaces within houses had become historically differentiated. His general conclusion shows a sense of space resembling Lefebvrean 'social space', rather than Bachelard's 'intimate space':

> A whole history remains to be written of spaces – which would at the same time be the history of powers (both these terms in the plural) – from the great strategies of geo-politics to the little tactics of the habitat, institutional architecture from the classroom to the design of hospitals, passing via economic and political installations. It is surprising how long the problem of space took to emerge as a historico-political problem. Space used to be either dismissed as belonging to 'nature' ... or else it was conceived as the residential site or field of expansion of peoples, of a culture, a language or a State ... Anchorage in a space is an economic–political form which needs to be studied in detail.[39]

For Foucault, space *is* power, and power is always spatially located somewhere within society: social relations of power infuse all spatial sites and concepts, from the micro- to the macro-level. As such, Foucault admitted in the course of an interview with French geographers: 'Geography must indeed lie at the heart of my concerns.'[40] Certainly, contemporary geographers have found Foucault's work very helpful in theorising space, place and power. Soja, for example, suggests that Foucault's theories show a 'spatialisation of history', where history is 'entwined with the social production of space'.[41]

Foucault's theory of spatiality shares much with Lefebvre's thinking: it is saturated with the social, and can be traced historically; it is political in nature; and it occurs in many different forms – body, room, house, institution, or political geography. Foucault, however, differs from Lefebvre in his rejection of the latter's Marxism with its privileging of relations of production. Instead Foucault substitutes his conception of *power*, arguing that the history of spaces is not the history of relations of production, but of relations of power.

A second difference from Lefebvre is Foucault's concept of heterotopia, found first in the short lecture from 1967 'Of other spaces' (*Des espaces autres*). For such a tantalisingly brief and cryptic piece, this text has occasioned much commentary.[42] The crucial value of the concept of heterotopia is its ability to connect material and metaphorical senses of space. It also presents a novel way for thinking about the relations between space and place in modernism.

'Of other spaces' commences by tracing a brief history of 'space', from the medieval to the present. The nineteenth century was, argues Foucault, obsessed with time and history, while the present epoch is captivated with spatial modes of thought, where the key conceptual tropes are those of simultaneity, juxtaposition, the network and the site (OS, p. 22). After a short discussion of Bachelard's work on the phenomenology of internal spaces, Foucault turns his attention to external spaces. Like Lefebvre's critique of abstract space, Foucault notes that 'the space in which we live ... is a heterogeneous space. In other words we do not live in a kind of void, inside of which we could place individuals and things' (OS, p. 23). Instead, Foucault defines the spaces in which we live as 'sets of relations that delineates sites' (p. 23). Indeed, a site is defined by its particular set of relations, an argument recalling Lefebvre's sense of the flows that interpenetrate any social space.

However, after some discussion of various types of site – such as those of transportation (the train, the street), relaxation (the café, the cinema, the beach) and of rest (the house, the bedroom, the bed) – Foucault outlines the type of site in which he is most interested. These are sites related to other sites, 'but in such a way as to suspect, neutralize, or invert the set of relations that they happen to designate, mirror, or reflect' (OS, p. 24). These sites of contradiction are distinguished into two sorts: utopias and heterotopias. A utopia is an *unreal* space, says Foucault, one that inverts an existing society or presents a perfect society. A heterotopia, however, is a *real* space that acts as a counter-site: it is 'a kind of effectively enacted utopia in which the real sites ... are simultaneously represented, contested, and inverted' (OS, p. 24). It is a place that is outside of all places, but which can be located in reality, unlike a utopia. Foucault cites the mirror as an instance of heterotopia. The mirror is actually located in reality; but the image of myself I see within it is located nowhere, in a virtual space. The mirror functions heterotopically because it contains both the real space and the unreal space simultaneously; or, more precisely, it functions as kind of 'counteraction' upon the person who gazes at the mirror. In order to see herself a person's gaze must pass through the reality of the actual glass of the mirror, and also through the unreality of the virtual image in the mirror. The real space of the mirror thus functions in a counter-real fashion. The important point is that heterotopia involves a sense of *movement* between the real and the unreal; it is thus a site defined by a process, the stress being upon the fact that it contests another site.

Foucault offers six principles for the study and description of heterotopias:

- Heterotopias are found in all human societies, but are extremely varied in their forms.[43]
- Heterotopias have a precise function for each historical society. An example given by Foucault is the role and placement of the cemetery. Until the end of the eighteenth century the cemetery was placed near a church, in the heart of the city. This symbolised the way in which bodily remains were regarded, linked to notions of the immortality of the soul and the resurrection of the body. During the nineteenth century cemeteries shifted to the edge of the city, and thence to the suburbs, signifying a loss of faith in doctrines of resurrection, as well as a sense of the decaying body as an object bearing illness (OS, p. 25).
- The heterotopic site is contradictory, and can juxtapose a number of different sites within it, a principle recalling Lefebvre's hypercomplex social space. Foucault's instances include the cinema and the theatre, rooms which contain other places, the screen or the stage, within them, and which open out into fictional other spaces.[44]
- Heterotopias are linked to 'heterochronies' – 'slices in time' – particular historical moments of rupture from traditional senses of time. Instances are the fairground, the festival and the holiday village, which are devoted to transitory moments in time. Joyce's *Ulysses* contains many instances of this interaction of heterotopia and heterochronia, for example, in the chapter 'Circe', set in the festival-world of the nighttown brothel.
- Heterotopias are generally linked to some system of opening and closing that means they are not freely accessible as public spaces. Entry is either by compulsion, as in a prison or military barracks; or one

has to fulfil certain criteria, such as religious rites for certain temples or hygienic gestures for baths or saunas.

- Heterotopias function and make sense only in relation to other forms of space. They function either as spaces of *illusion* that show up real spaces as more illusory even than fantasy spaces; or they operate by *compensation*, creating a 'space that is other, another real space, as perfect, meticulous, as well arranged as ours is messy, ill constructed, and jumbled' (OS, p. 27). Heterotopias of this second kind – 'absolutely perfect other places' (p. 27) – might include certain colonies, says Foucault, such as Jesuit colonies in South America, or Puritan colonies in seventeenth-century America. Foucault should not be interpreted as suggesting that colonies were, in reality, perfect places; rather that their conception was guided by this kind of social and spatial goal. Thus these colonial villages were constructed around specific regular formations – the grid, the cross at right angles, the central location of the church. Daily life was also regulated and regimented according to a strict timetable. This example shows how Foucault's conception of heterotopia is not necessarily of a positive place of freedom or escape. Heterotopias, in Kevin Hetherington's succinct summary, are simply spaces for an 'alternate ordering' of modernity: they 'organize a bit of the social world in a way different to that which surrounds them'.[45]

Clearly, not all of these six principles are fully theorised by Foucault, and some are left protean and not a little contradictory. But there is a richness about the vision of space offered here, particularly when applied to modernism. This is demonstrated in the frustratingly brief final paragraph of the lecture, where Foucault describes one more instance of heterotopia: a boat, which is

> a floating piece of space, a place without a place, that exists by itself, that is closed in on itself and at the same time is given over to the infinity of the sea and that, from port to port, from tack to tack, from brothel to brothel, it goes as far as the colonies in search of the most precious treasures they conceal in their gardens ... the boat has not only been for our civilization, from the sixteenth century until the present, the great instrument of economic development ... but has simultaneously been the greatest reserve of the imagination. The ship is the heterotopia par excellence. In civilizations without boats, dreams dry up, espionage takes the place of adventure, and the police take the place of pirates. (OS, p. 27)

This cryptic portrait of the boat as heterotopia indicates a fertile link to literary dreams of 'other spaces'. We need think only of the central role played, for example, by boats in Conrad's narratives to understand how we might use heterotopia as a tool for revealing complex modernist geographies. *Heart of Darkness*, for example, not only concerns the central journey of a boat into a colony, but is entirely narrated aboard a boat moored on the Thames. The story that emerges on this 'place without a place' functions in relation to the nearby modern city in a strikingly heterotopic fashion: its tale of the savagery of Kurtz in Africa brutally exposes the facade of the civilised city of London. The rational modernity of the imperial city space is shown up as illusion by the violence of the story narrated on the boat. *Heart of Darkness* constantly stresses the relations between the excesses of imperial trade in Africa and the order of the metropolis, and the boat's mediating function here is precisely heterotopic. For heterotopia incorporates diverse places, moving between and across them, and in so doing reveals the processes that link together different kinds of *topoi*.[46] The heterotopic boat in *Heart of Darkness* is a thus a contradictory site, fixed by its anchor, yet moving in its narrative space between the Congo, Brussels and London to illuminate the interconnections between the European metropolis and its imperial domains.[47]

There seem, then, to be productive ways in which Foucault's heterotopia can be used in the reading of modernism. Foucault himself, in his only other explicit use of the term, linked heterotopia to literature in *The Order of Things*, published in 1966, a year before 'Of other spaces'. Prompted by an essay of Borges that mocks attempts to tabulate human knowledge, Foucault uses heterotopia to describe a form of writing that undermines the idea of such an ordering of knowledge. This kind of disorder is not just about placing incongruous items together, as in the Surrealist image cited by Foucault, of combining together an umbrella and a sewing-machine on an operating table.[48] Rather it involves doing away with the operating table itself, or the *site* upon which strange objects are grouped together:

> I mean the disorder in which fragments of a large number of possible orders glitter separately in the dimension, without law or geometry, of the *heteroclite*; in such a state, things are 'laid', 'placed', 'arranged' in sites so very different from one another that it is impossible to find a place of residence for them, to define a common locus beneath them all.[49]

Foucault distinguishes this groundless writing from utopias, which offer a consolation in their untroubled image of a perfect order. Heterotopias are considerably more disturbing

> because they secretly undermine language, because they make it impossible to name this *and* that, because they shatter or tangle common names, because they destroy 'syntax' in advance, and not only the syntax with which we construct sentences but also that less apparent syntax which causes words and things (next to and also opposite one another) to 'hold together' ... [H]eterotopias ... dessicate speech, stop words in their tracks, contest the very possibility of grammar at its source; they dissolve our myths and sterilize the lyricism of our sentences.[50]

This is clearly a description of a form of avant-garde writing found throughout modernism, associated in Foucault's mind with writers such as Bataille and Blanchot.[51] As Foucault sums it up, this writing shows a linkage of *aphasia* (speech disorder) with *atopia* (disturbances of place). But it can easily be applied to the modernist style of Joyce in *Ulysses* or more widely in *Finnegans Wake*, the disrupted syntax of much of Gertrude Stein, or the patchwork texture of Ezra Pound's *Cantos*.

Significantly, in 'Of other spaces', Foucault does not return to this strictly literary heterotopia, because he wishes to extend this metaphorical sense of space to heterotopias found in actual spaces. This seems the most important point to glean from Foucault's comments on heterotopias for the study of modernism, that the metaphorical subversions of heterotopic writing be brought into contact with the actual sites and countersites of modernity. Together they articulate an interpretation of modernism as a set of responses to changes in the material spaces of modernity, shown in, for instance, Forster's image of metropolitan suburbs in *Howards End* or Imagist poems set on underground trains. Equally, however, this approach to modernism analyses metaphorical spaces in the text, such as the space of the body in *Ulysses*, or the interior space of consciousness in Rhys and Woolf. Literary texts, in another sense, combine both metaphorical and material space if we focus upon their formal properties. The disorientating heterotopia of a modernist narrative might be directly indebted to urban space; experiments with typography and line spacing in modernist poetry could be linked to the emergence of heterotopic sites in modernist cities. Here the material form of the text is a transformation of some specific external space; turning, for example, the streets of Dublin into a meandering narrative in *Ulysses*, such that we read the twists and turns of meaning as an embodiment of urban space. If one function of heterotopia is, in Genocchio's words, to 'inscribe instability into a given spatial order',[52] then we should look for those moments in a text in which linguistic or semantic instability is associated with a certain site or location in order to find modernist heterotopias.

Use of the concept of heterotopia should also recall Foucault's comments in the 1970s upon the significance of power in space and geography, a topic not directly addressed in the earlier discussion of heterotopia. Power is implied since heterotopia inverts and contests real sites, and makes sense only in relation to some other real space in society. Certain commentators have interpreted heterotopias as simply sites of resistance to the dominant ordering of socio-spatiality, found in marginal places and locations.[53] However, Foucault's conception of power as a set of relations rather than an object one could possess suggests that heterotopias cannot be labelled as inherently sites of resistance. Heterotopias are not sites of absolute freedom or places where marginal groups always resist power; as Hetherington observes, the important point is 'not the spaces themselves but what they perform in relation to other sites'.[54] Foucault's list of heterotopias includes prisons as well as gardens, colonies as well as ships. The heterotopic ship in Conrad's *Heart of Darkness* may function as a critique of imperial exploitation, but it is still a ship registered and engaged in those very imperialist practices. We should be aware, then, when reading modernist heterotopias to indicate the ambivalent strategies of power informing their spatial practices. An example of the potentially ambivalent character of heterotopia might be Forster's use of the house and garden at Howards End. Is it the site of an alternative ordering of modernity or merely of a nostalgic escape from modernity? Perhaps the motorcar in the novel is another heterotopia: a real site but one which will not stay put, a 'placeless place' that constantly unsettles an acceptable spatial ordering of modernity. The disruptions of Forster's narrative form and style when describing the motorcar also indicate the linguistic instability of a textual heterotopia. For Forster, however, this is quite clearly a negative heterotopia.

The concept of heterotopia thus represents a fluid sense of social space, and the processes to which space is subject. It shows 'place' to be, contra Heidegger's dwelling, a site of potential instability. It is, therefore, a concept which connects material and metaphorical spaces in the literary text in new and illuminating ways, showing how the formal practices and spatial form of the modernist text should be read in conjunction with a wider understanding of the historical geography of modernity.

De Certeau and the syntax of space

In *The Practice of Everyday Life* (1984 (hereafter *PE*)) Michel de Certeau outlines a theory of how people contest the imposition of various forms of power in their daily lives by specific 'ways of using the products imposed by a dominant social order' (*PE*, p. xiii).[55] De Certeau draws upon Foucault's analysis in *Discipline and Punish* of the creation in modern societies of a 'disciplinary power' that seeps through the social body in an anonymous fashion, producing subjects who are moulded by a diffuse mechanism of social obligation. De Certeau's aim is to demonstrate the other side of this disciplinary matrix: the forms of resistance to such powers embodied not in grand political strategies or projects, but in the quotidian activities of the ordinary person. De Certeau's book studies the 'ways of operating [that] manipulate the mechanisms of discipline and conform to them only in order to evade them' (*PE*, p. xiv).

One of the central insights found in de Certeau is his focus upon the sites of such resistances: 'these "ways of operating" constitute the innumerable practices by means of which users reappropriate the space organized by techniques of sociocultural production' (p. xiv). De Certeau takes this insight from the stress in Foucault's work upon the locations where disciplinary power is instigated, most famously in the example of the prison panopticon. De Certeau's examples of this contestation of space centre upon walking in the city, travelling by rail and the idea of 'spatial stories'. There are two points in de Certeau's work that are pertinent for this discussion of modernism. The first is his insistence upon the relationships of power that suffuse people's occupations and use of space, an argument that extends Foucault's putative link between the history of spaces and the history of powers. Secondly, de Certeau's specific notion of 'spatial stories' as a practice of urban life emphasises the combination of material and metaphoric spaces. The concept of 'spatial stories' thus refines the analogy between language and practice by arguing that '[e]very story is a travel story – a spatial practice' and that all stories 'traverse and organize places; they select and link them together; they make sentences and itineraries out of them' (*PE*, p. 115).[56] In a sense de Certeau's spatial stories connect Lefebvre's social space and the formal practices of the literary text.

This link between space and narrative illuminates a series of points about the linguistic form of modernist narratives. One is the way that modernist works interrogate the quotidian spaces of modernity: the city, the street, the room, the house, and so on. Another concerns the implications of de Certeau's claim that 'narrative structures have the status of spatial syntaxes' (p. 115). Like Foucault's textual heterotopias, the notion of 'spatial stories' offers a method that focuses upon more than simply how particular modernist texts represent particular spaces: it allows us to outline the formal strategies by which spaces are represented and to understand the relation between text and space in a more interactive fashion. The spaces of modernity alter and transform the literary space of early twentieth-century writing; while the peculiar spatial stories told in the literary texts of modernism shape the ways in which we view and understand modernity itself.

Two corresponding sets of distinctions introduced by de Certeau provide guidance for this approach: a demarcation between *space* and *place*; and a related division between *tours* and *maps*. For de Certeau, perhaps following Heidegger, a place 'implies an indication of stability' and is a location where 'elements are distributed in relationships of coexistence' (*PE*, p. 117). Two things cannot occupy the same place: elements can only exist beside one another, each situated in its 'proper' location. De Certeau uses 'proper' to mean the official and legitimised use to which a place or activity belongs. A space, however, is based not on stability but on direction, movement and velocity:

> space is composed of intersections of mobile elements. It is in a sense actuated by the ensemble of movements deployed within it. Space occurs as the effect produced by the operations that orient it, situate it, temporalize it, and make it function in a polyvalent unity of conflictual programs or contractual proximities. (*PE*, p. 117)

This sense of space as produced by the intersection of different elements owes much to Lefebvre's idea of 'the hypercomplexity of social space'. Space, writes de Certeau, 'is a practiced [*sic*] place' (*PE*, p. 117), it is like the meaning of a word actually being spoken rather than its 'proper' meaning as found in a dictionary. De Certeau's example is that of a city street planned in a geometrical fashion. This place is 'transformed into a space by walkers' in much the same way that 'an act of reading is the space produced by the practice of a particular place: a written text, i.e., a place constituted by a system of signs' (p. 117). Places, argues de Certeau, are always determined by a focus upon fixity, or what he calls 'the *being-there* of something dead', so that an inert object serves as the foundation of place. Spaces, however, are determined by *operations* attributable to historical subjects rather than lifeless bodies: 'a movement always seems to condition the production of a space and to associate it with a history' (p. 118). 'Place' thus resembles Lefebvre's official representation of space, while 'space' is closer to a combination of Lefebvre's representational space and spatial practice.

Stories constantly oscillate around these two poles, transforming spaces into places, and places into spaces. For example, place becomes space when an inert object, 'a table, a forest, a person that plays a certain role in the environment' (*PE*, p. 118), emerges from a stable location and transforms it by narrative action into a space. De Certeau is careful to suggest that these are not unchanging binary terms, since places and spaces are constantly being transfigured into one another in the play of narrative. He also argues that it might be possible to produce a typology of the ways in which stories enact either an '*identification of places*' or an '*actualization of spaces*' (p. 118).

De Certeau's distinction between space and place can be illustrated by considering how modernist narratives broadly differ from those of realism. De Certeau describes a kind of story in which we witness 'the putting in place of an immobile and stone-like order' and where 'nothing moves except for discourse itself, which, like a camera panning over a scene, moves over the whole panorama' (*PE*, p. 118). This seems an apt description of many features of the realist novel of the nineteenth century, noticeable in an extended passage such as the opening of chapter 2 of Thomas Hardy's *Tess of the D'Urbervilles* (1891):

> The village of Marlott lay amid the north-eastern undulations of the beautiful Vale of Blakemore or Blackmoor aforesaid, an engirdled and secluded region, for the most part untrodden as yet by tourist or landscape-painter, though within a four hours' journey from London.
>
> It is a vale whose acquaintance is best made by viewing it from the summit of the hills that surround it – except perhaps during the droughts of summer. An unguided ramble into its recesses in bad weather is apt to engender dissatisfaction with its narrow, tortuous, and miry ways.
>
> This fertile and sheltered tract of country, in which the fields are never brown and the springs never dry, is bounded on the south by the bold chalk ridge that embraces the prominences of Hambledon Hill, Bulbarrow, Nettlecombe-Tout, Dogbury, High Stoy, and Bubb Down. The traveller from the coast, who, after plodding northward for a score of miles over calcareous downs and corn-lands, suddenly reaches the verge of one of these escarpments, is surprised and delighted to behold, extended like a map beneath him, a country differing absolutely from that which he has passed through. Behind him the hills are open, the sun blazes down upon fields so large as to give an unenclosed character to the landscape, the lanes are white, the hedges low and plashed, the atmosphere is colourless. Here, in the valley, the world seems to be constructed upon a smaller and more delicate scale; the fields are mere paddocks, so reduced that from this height their hedgerows appear a network of dark green threads overspreading the paler green of the

> grass. The atmosphere beneath is languorous, and is so tinged with azure that what artists call the middle distance partakes also of that hue, while the horizon beyond is of the deepest ultramarine. Arable lands are few and limited; with but slight exceptions the prospect is a broad rich mass of grass and trees, mantling minor hills and dales within the major. Such is the Vale of Blackmoor.[57]

This is a beautiful evocation of place drawing upon an explicitly topographical style of visual description. Hardy suggests comprehending the landscape from above, like a painter viewing the land 'extended like a map beneath him'. Attention is drawn to other visual markers of place, such as scale (the valley seems smaller and of a more 'delicate scale') and shape (the hedgerows that appear as 'a network of dark green threads'). We are also located precisely in terms of the names of surrounding villages, by reference to compass directions and by the distance from London. Hardy's discourse sweeps over this landscape, quietly yielding up the 'secluded region' for the eye of the reader, utilising what Lefebvre terms a 'logic of visualisation' (*PS*, p. 287) to identify and pin in narrative form the nature of Marlott. But de Certeau's sense of space is absent here, precisely because there is no actualization of the landscape by any acting or moving subject.

The passage also illustrates another feature of de Certeau's 'identification of place': reliance on a mode of discourse he terms the *map*. This is juxtaposed with the *tour*, an experiential discourse associated with the 'actualization of space'. Map discourses order precisely where elements or features occur; the hypercomplexity of social space is constrained by a visual discourse which presents a 'tableau' or 'knowledge of an order of places', and finding its apogée in the conventional map with its 'plane projection totalising observations' (*PE*, p. 119). Hardy's text maps the space of the countryside around Marlott, transforming it into the immobility of a known and visually perceived place. Tour discourses refuse to present a visual tableau and are, instead, rooted in 'spatializing actions' that 'organize *movements*' (p. 119). The difference is between a discourse that lists where sites are located ('The girl's room is next to the kitchen, opposite the bathroom') and one that describes a location through a set of actions ('You enter the hallway, go along, you turn right, and then go across the room'). Though Hardy introduces an anonymous 'traveller from the coast' to bestow some human presence upon the scene, this person soon stands still and disappears from the landscape. The use of place names only intensifies the cartographic quality of this writing.

For de Certeau, most narratives combine both map and tour discourses, although one mode tends to dominate. The tour is the primary form of spatial discourse because it is connected with the actions of human subjects through space; these activities are subsequently codified into mapped places. The growth of the scientific discourse of modern mapping has, de Certeau argues, over a period of centuries gradually suppressed the role of tour itineraries.[58] The map 'colonizes space' and eliminates the movements of subjects who had initially produced these very spaces:

> The map, a totalizing stage on which elements of diverse origin are brought together to form the tableau of a 'state' of geographical knowledge, pushes away into its prehistory or into its posterity, as if into the wings, the operations of which it is the result or the necessary condition. It remains alone on the stage. The tour describers have disappeared ... maps, constituted as proper places in which to exhibit the products of knowledge, form tables of legible results. Stories about space exhibit on the contrary the operations that allow it, within a constraining and non-'proper' place, to mingle its elements anyway. (*PE*, p. 121)

Hardy's discourse identifying place is indeed one in which the 'tour describers have disappeared'; everything has its proper position in this narrative, and the land, the sea, the city and the hills do not intermingle or overlap in this mapped place.[59]

In contrast we could point to the way in which a modernist text like Joyce's *Ulysses* is much more likely to employ a tour form of discourse, in which the actualisation of space is privileged over the mapping of place. The following passage is from the start of the 'Cyclops' episode:

> By lorries along sir John Rogerson's quay Mr Bloom walked soberly, past Windmill lane, Leask's the linseed crusher's , the postal telegraph office. Could have given that address too. And past the sailors' home. He turned from the morning noises of the quayside and walked through Lime street. By Brady's cottages a boy for the skins lolled, his bucket of offal linked, smoking a chewed fagbutt. A smaller girl with scars of eczema on her forehead eyed him, listlessly holding her battered caskhoop. Tell him if he smokes he won' t grow. O let him! His life isn't such a bed of roses! Waiting outside pubs to bring da home. Come home to ma, da. Slack hour: won't be many there. He crossed Townsend street, passed the frowning face of Bethel. El, yes: house of: Aleph, Beth. And past Nichols' the undertaker's. At eleven it is. Time enough. (*U*, p. 68[60])

Because of Joyce's scrupulous attention to Dublin we might take this passage to exhibit an mapping of urban place. But the references to streets and buildings, in de Certeau's terms, actualise space rather than identify place, since the pedestrian movements of Leopold Bloom produce them. Bloom's perambulation through the city is a vast tour itinerary since, stylistically, the passage is full of the verbs and terms of movement: Bloom 'walked soberly', he goes 'past' buildings, he 'turned' and 'walked through [Lime street]', and 'crossed Townsend street'. More than this, however, we notice how the use of interior monologue is another technique for the actualisation of space. Joyce describes how Bloom passes the Postal Telegraph Office, but then shifts the discourse from this exterior picture to the interior musings of Bloom about using the Post Office address for illicit correspondence from a possible lover. The sight of the boy smoking prompts an inner dialogue about social disadvantage versus moral guidance on health, and an imagined story of why the boy is waiting outside the pub. Moving past the undertakers serves to remind Bloom of his coming engagement at a funeral. Exterior places therefore serve as the sites for the construction of spatial stories. Or, as de Certeau notes, 'the street geometrically defined by urban planning is transformed into a space by walkers' (*PE*, p. 117).

The passage also demonstrates de Certeau's point that spatial stories contain both space and place, tour and map in dialogue. Joyce famously wrote *Ulysses* with the aid of a Dublin street map, as well as Thom's Directory of Dublin Businesses, and there are parts of the novel that are clearly indebted to an identification of place through a kind of semi-cartographic discourse. The 'paralysed' place that Joyce perceived as Dublin is, however, transformed in his text into the bustling space of a thousand tours and stories; thus the mapping of Dublin places – the street names, the tram destinations, and the buildings – prompts a spiralling of spatial stories. *Ulysses* displays a perpetual encounter between the stability of the realist novel's map and the myriad movements and tours of Dublin characters through space.

De Certeau's work on 'spatial stories' thus emphasises how the distinction between space and place should not be reified into an absolute division. Modernism often oscillates between the two discourses of tour and map, as subsequent chapters show. De Certeau's book is helpful, then, for focusing upon the interaction of space and place, for its stress upon relations of power derived from Foucault, and for his linkage of language and spatiality. To conclude this survey or tour of spatial theorists David Harvey's account of modernism and modernity in *The Condition of Postmodernity* is considered. This is a work which draws upon a number of the writers already discussed, specifically Lefebvre, and suggests a number of congruences between certain theories of space and place and the formal literary innovations of modernism.

David Harvey: modernity and postmodernity

Harvey's *The Condition of Postmodernity* has been very influential in discussions of postmodernist culture, particularly for its attempt to understand the geographical and spatial character of postmodernity. The book has been central in introducing the concepts of space and geography into cultural and literary studies, and so it is helpful to discuss how Harvey concieves of the relation between space and place in modernism and modernity.

Harvey's analysis of postmodernity relies upon an argument about the contours of modernity and modernism as cultural experiences that are transformed, but not jettisoned, in the present condition of postmodernity. Modernity is characterised by an experience of 'time–space compression' in capitalist societies, defined as a 'speed-up in the pace of life, while so overcoming spatial barriers that the world sometimes seems to collapse inwards upon us'.[61] This powerful experience of compression in our quotidian grasp of time and space is 'challenging, exciting, stressful, and sometimes deeply troubling, capable of sparking ... a diversity of social, cultural, and political responses'.[62] Harvey analyses how modernity, from the middle of the nineteenth century onwards, experienced just such an ambivalent 'time–space compression', a process distinguishing that epoch from the earlier conceptions of time and space initiated in the Enlightenment. To comprehend time–space compression Harvey draws upon Marshall Berman's tripartite division between modernisation, modernity and modernism.[63] Time–space compression begins with capitalist modernisation, the need, first analysed by Marx, to constantly revolutionise economic production by the use of new technologies and practices in order to extend and maintain profits. This produces a set of social, psychological, and cultural responses to the accelerated nature of everyday life undergoing technological and political modernisation; the accumulated experience of these changes is termed modernity. Finally, modernism is, in Harvey's words, 'a troubled and fluctuating aesthetic response to conditions of modernity produced by a particular process of modernization.'[64]

One of the major achievements of Harvey is to demonstrate how time–space compression interweaves these three levels of modernisation, modernity and modernism. The chapter 'Time–space compression and the rise of modernism as a cultural force' contains an important discussion of the emergence of modernist literature in relation to conceptualisations of space and place. Harvey follows de Certeau in privileging space over place, finding a strain in modernism that reacted to the anxieties produced by time–space compression by encouraging an 'identification of place' that was ultimately conservative, nostalgic, and linked to reactionary political programmes such as Fascism. Drawing upon Nietzsche and Heidegger, Harvey associates place with a conception of *being*, while space is linked to a notion of *becoming*.[65]

For Harvey the birth of modernism after 1848 commences a protest against the practical and theoretical rationalisation of space and time in Enlightenment thought. Due to the internationalising of capital, European space became more unified towards the latter half of the nineteenth century. Modernisation, in the form of technological inventions such as the railway, the telegraph, steam-shipping and the radio, all helped the way in which capitalism engaged in a 'massive long-term investment in the conquest of space'.[66] Linked to this technological take-off was the development of imperialism at the end of the century. Up to the outbreak of the First World War 'the world's spaces were deterritorialized, stripped of their preceding significations, and then reterritorialized according to the convenience of colonial and imperial administrations ... The map of domination of the world's spaces changed out of all recognition between 1850 and 1914.'[67] Such massive transformations resulted in immensely original experiences of everyday life, a new sense of modernity, where a Londoner could consume a product – the peach, for example, that Eliot's J. Alfred Prufrock may or may not dare eat – shipped across thousands of miles from a colony owned and ruled from one's metropolitan centre. Modern time and space now *felt* different, because, for example, one could travel by train at hitherto unimagined speeds, converting vast stretches of space into mere blips on the clock.[68]

This process, whereby the earth's surface is resignified, is brilliantly shown in Conrad's *Heart of Darkness*. Marlow recalls his youthful pleasure in looking at maps and dreaming of the 'glories of exploration' of the 'many blank spaces on the earth'.[69] As an adult, however, those spaces were no longer empty on the map. The map of Africa, 'the biggest – the most blank, so to speak', had altered significantly: 'It had got filled since my boyhood with rivers and lakes and names. It had ceased to be a blank space of delightful mystery ... It had become a place of darkness.'[70] Imperialist expansion is perceived as a process of converting unknowable – to Western eyes at least – spaces into cognised and mapped places. In de Certeau's terms, Marlow's regret is for the transformation of spaces into identified places, for a tour that is reduced to a mere mapping.

Harvey sketches three ways in which modernism produced a set of correspondences to these massive revolutions in modernisation and modernity. First, Realist narrative, the established norm of the novel, came under duress towards the end of the century. Realism relied upon a linear narrative schema that seemed unsustainable in a world where events did not necessarily occur in consecutive time. If, simultaneously, a trader in America could talk business, transacting a shipment of peaches, on the telephone with a banker in London, then it seemed as if two spatially disjunct events could occur at the same time. Thus Realist narrative 'structures were inconsistent with a reality in which two events in quite different spaces occurring at the same time could so intersect as to change how the world worked'.[71] The narrative form of Joyce's 'Wandering Rocks' chapter of *Ulysses*, where a number of simultaneous events are represented from eighteen different spatial perspectives, is an example of literary modernism trying to capture the reality of this time–space compression of modernity.

Modernism's second response to time–space compression, according to Harvey, was the idea of an interiority that existed outside of the rationalised exterior spaces of a modernity which eradicated local differences between various groups and people. Harvey argues that the space of 'the body, of consciousness, of the psyche' had been repressed by 'the absolute suppositions of Enlightenment thought'. Now, however, 'as a consequence of psychological and philosophical findings' such interiorised spaces could 'be liberated only through the rational organization of exterior space and time'.[72] Harvey's argument is here indebted to Lefebvre: modernism's 'representational spaces' of the body or the psyche are reactions to the abstract and technologised 'representation of space' in modernity. Much modernist writing explores these interiorised spaces, not just as a result of innovations in literary history, but as a keenly felt response to fundamental alterations in the spatial history of modernity in the early twentieth century. This is a significant point for interpreting modernism: that the bodily space of a Leopold Bloom or the psychic space of a Clarissa Dalloway must be understood in relation to a wider understanding of social space and geographical history.

The third feature of modernism's reaction to time–space compression, for Harvey, focuses upon the meanings imbued to particular places in literary texts. In a period when space was being conquered by time, and an internationalist consciousness of various sorts was being fostered, there was also a growing sense of the importance of locale and the uniqueness of place. Harvey finds the rise of this 'identification of place' in features such as the rise of vernacular architecture or the recurrent motif of *chinoiserie* in French painting at the turn of the century. In much of Pound's Imagist poetry, for example, we find reference to Chinese themes and, later in his career, the use of a non-Western ideogrammatic language.[73] Thus, writes Harvey, the 'identity of place was reaffirmed in the midst of the growing abstractions of space'.[74] Though modernism is most often associated with a certain international outlook, such as the cosmopolitanism to which Forster frequently refers in *Howards End*, it is clear that many modernist texts draw upon and reproduce a mythology of individual places: Eliot, Woolf and Dorothy Richardson on London; Joyce on Dublin; the Paris of Djuna Barnes and Rhys.

Harvey's attitude to this focus upon place is, ultimately, a critical one, since he views it as a trend in which *being*, 'the spatialisation of time', is privileged over *becoming* 'the annihilation of space by time'.[75] Given Harvey's allegiance to Marxism, with its profound interest in the role of historical *becoming* in the progressive development of society, it is not surprising that he repeats a general claim made about modernism by many Marxist critics: that it attempts to evade or reject history in favour of a spatialised politics.[76] Harvey, therefore, is uneasy with this localised dimension of modernism:

> Modernism, seen as a whole, explored the dialectic of place versus space, of present versus past, in a variety of ways. While celebrating universality and the collapse of spatial barriers, it also explored new meanings for space and place in ways that tacitly reinforced local identity. By enhancing links between place and the social sense of personal and communal identity, this facet of modernism was bound, to some degree, to entail the aestheticization of local, regional, or national politics.[77]

Harvey's dismissal of modernism's engagement with place is, however, overstated and rather one-dimensional. In the writings of Joyce and Rhys in particular we see a modernism engaged in a dialectic between space and place in such a way that the stress upon local identities does not necessarily result, as Harvey suggests, in an aestheticised and reactionary politics of the nation. Harvey's argument here is bluntly formulated and is too locked into the binary categories he employs (being–becoming, place–space) to be able to adequately discern the more nuanced uses of space and geography in modernist literature. If we recall the similar argument about space and place made by de Certeau, we can say that identification of place and actualisation of space are the twin poles around which the spatial stories of modernism operate. Foucault's notion of heterotopia, linking material and metaphorical spaces, also offers a more fluid sense of space, place and power than is found in this aspect of Harvey's account of modernism. Harvey's work is illuminating, however, in its general claim that 'time–space compression' was a significant factor in the aesthetic trajectory of modernism.

Conclusion

This chapter has argued for the validity of thinking modernism in terms of space and place, and has outlined a number of ideas drawn from a variety of thinkers to help formulate this critical literary geography of modernism. These theories inform the readings of the literary texts found in the rest of the book. Therefore, I do not return to them in any detailed fashion later, hoping the reader might return to this chapter if at all puzzled by any of the ideas expressed later.

From Heidegger I have taken the central, but problematic, distinction between space and place, an opposition crucial for many other spatial theorists. Bachelard's work on the poetics of space indicated the importance of interiors and the intimate space of the body, although his discourse on place contained a number of limitations. In Lefebvre's complex work *The Production of Space* there are a number of stimulating ideas about 'social space', primarily that of the distinction between 'representations of space', 'representational spaces' and 'spatial practices'. Lefebvre's work also introduced questions of power and politics into the discussion of space, topics taken up again in Foucault's work on heterotopia. Perhaps Foucault's most significant insight in his writings on heterotopia, and elsewhere, was his suggestion that material and metaphorical senses of space must be combined. This is a point developed in de Certeau's notion of spatial stories, and in his contrast between the tour and the map.

Modernism, then, as this book explores in the following chapters, was powerfully obsessed with the relations between space and place, as Harvey argues. But in order to understand the complexity of this phenomenon we must interpret modernist images of space and place with an eye on the form and styles of the writing itself. Any critical literary geography must, therefore, trace the *textual space* of modernism: that is, how the spatial form of the literary text is inextricably linked to the histories of social space, and how we find a combination of metaphoric and material spaces represented in modernist writings. Chapter 2 begins this movement through the spaces of modernity with a text, Forster's *Howards End*, published in 1910, the year for Virginia Woolf when 'human character changed', and the year in which, suggested Henri Lefebvre, 'a certain space was shattered' (*PS*, p. 25).[78] It is also a text with two central images representing the relations between place and space: the house at Howards End, and the motorcar that zips through the new spaces of modernity.

Notes

1 David Harvey, *The Condition of Postmodernity* (Oxford, Blackwell, 1990), p. 24.
2 Malcolm Bradbury and Alan McFarlane, eds, *Modernism: A Guide to European Literature 1890–1930* (Harmondsworth, Penguin, 1976).
3 The geographical writing on place is extensive; see, for example, Yi-Fu Tuan, 'Space and place: humanistic perspective' (1974) reprinted in John Agnew, David N. Livingstone and Alisdair Rogers, eds, *Human Geography: An Essential Anthology* (Oxford, Blackwell, 1996); John A. Agnew and James S. Duncan, eds, *The Power of Place: Bringing Together Geographical and Sociological Imaginations* (Boston, MA, Unwin Hyman, 1989); Michael Keith and Steve Pile, eds, *Place and the Politics of Identity* (London, Routledge, 1993); Doreen Massey and Pat Jess, eds, *A Place in the World? Places, Cultures and Globalization* (Oxford, Oxford University Press, 1995).
4 Martin Heidegger, 'Building, dwelling, thinking' in *Poetry, Language, Thought*, trans. Albert Hofstadter (New York, New Directions, 1971) p. 147.
5 Heidegger, 'Building', p. 156.
6 Heidegger, 'Building', p. 157.
7 Heidegger, 'Building', p. 157.
8 Heidegger, 'Building', p. 160.
9 For this criticism see Doreen Massey 'Power-geometry and a progressive sense of place', in Jon Bird, Barry Curtis, Tim Putnam, George Robertson and Lisa Tickner, eds, *Mapping the Futures: Local Cultures, Global Change* (London, Routledge, 1993).
10 For example, Seamus Heaney's essay 'The sense of place' appears to draw upon Heidegger when the nature of one's parochial location in a geography is praised for its refusal to engage with any wider social or historical geography: 'The sense of place' in *Preoccupations: Selected Prose 1969–1978* (London, Faber, 1980).
11 Gaston Bachelard, *The Poetics of Space*, trans. Maria Jolas (Boston, MA, Beacon Press, 1969), p. 7.
12 Bachelard, *Poetics*, p. 3.
13 Bachelard, *Poetics*, pp. 4, 5.
14 Bachelard, *Poetics*, p. xxxii.
15 Bachelard, *Poetics*, p. xxxii.

16 Edward S. Casey, *The Fate of Place: A Philosophical History* (Berkeley and Los Angeles, University of California Press, 1997), p. 291.
17 Casey, *Fate of Place*, p. 292.
18 Bachelard, *Poetics*, p. 14.
19 Bachelard, *Poetics*, p. 15.
20 Bachelard, *Poetics*, p. xxxi.
21 Charlotte Perkins Gilman, *The Charlotte Perkins Gilman Reader: 'The Yellow Wallpaper' and Other Fiction* (London, Women's Press, 1979).
22 Though Casey argues that Bachelard's discussion of 'intimate immensity' dissolves the interior–exterior opposition, so that we feel 'outerness inside'; see Casey, *Fate of Place*, p. 295.
23 Henri Lefebvre, *The Production of Space*, trans. Donald Nicholson-Smith (Oxford, Blackwell, 1991). For a discussion of the impact of Lefebvre's work see Edward W. Soja, *Thirdspace: Journeys to Los Angeles and Other Real-and-Imagined Places* (Oxford, Blackwell, 1996), pp. 25–82.
24 Henri Lefebvre, 'Reflections on the politics of space', *Antipode* 8:2 (May 1976), 31.
25 Soja, *Postmodern Geographies*, pp. 57–8.
26 See Mikhail Bakhtin, 'Discourse in the novel' in *The Dialogic Imagination: Four Essays*, ed. Michael Holquist, trans. Caryl Emerson and Michael Holquist (Austin, University of Texas Press, 1981).
27 Lefebvre criticises Heidegger for his neglect of space in favour of time, and for his use of a concept of absolute space – space as an empty container – without a history of its production (*PS*, pp. 121–2).
28 Here I am using Harvey's translations of Lefebvre's descriptions of his three terms. In the English-language version of Lefebvre the following terms are used: spatial practices are *perceived* (the experienced); representations of space are *conceived* (the perceived); and representational spaces (what Harvey calls spaces of representation) are *lived* (the imagined). See Harvey, *Postmodernity*, pp. 218–9, and the grid on pp. 220–1.
29 One example, however, is the development of the Italian Renaissance hill-towns of Tuscany, which Lefebvre links to the new space that was conceptualised by writers, architects and painters by means of the theory of perspective (*PS*, p. 79).
30 For the distinction between dominant space and appropriated space see *PS*, pp. 164–8.
31 Surprisingly, for its date of initial publication, there is also much in *The Production of Space* on corporeal space, and on gender and sexuality. For a consideration of the significance of the body in Lefebvre's work see Steve Pile, *The Body and the City: Psychoanalysis, Space and Subjectivity* (London and New York, Routledge, 1996), pp. 145–69.
32 For the impact of Foucault on geography see in particular C. Philo, 'Foucault's geography', *Environment and Planning D: Society and Space* 10 (1992), 137–61; and Soja, *Thirdspace*, pp. 145–63.

33 Michel Foucault, 'Of other spaces', *Diacritics* 16: 1 (spring 1986).

34 Edward Said, Interview in Imre Salusinszky, ed., *Criticism in Society* (New York and London, Methuen, 1987), p. 137; John Marks, 'A new image of thought', *New Formations* 25 (summer 1995), 69.

35 Michel Foucault, 'Questions on geography' in *Power/Knowledge: Selected Interviews and Other Writings 1972–1977* (London, Harvester Wheatsheaf, 1980), pp. 69–70.

36 Deleuze describes the *dispositif* as a concrete social apparatus, a 'tangle' or 'multilinear ensemble', that combines knowledge and power in a variety of spatial tropes and material locations. See Gilles Deleuze, 'What is a *dispositif?*' in *Michel Foucault, Philosopher*, trans. Tim Armstrong (Hemel Hempstead, Harvester Wheatsheaf, 1992), pp. 159–60.

37 See Michel Foucault, *Discipline and Punish: The Birth of the Prison*, trans. A. M. Sheridan-Smith (Harmondsworth, Penguin, 1977), pp. 195–228.

38 Deleuze comments that although most commentators view Foucault's conceptualisation of space 'in terms of cartography, topography, or geology' it is, in fact, three-dimensional. See Gilles Deleuze, 'Un nouvel archiviste', quoted in Rob Shields, *Places on the Margin: Alternative Geographies of Modernity* (London, Routledge, 1991), p. 40).

39 Foucault, 'The eye of power' in *Power/Knowledge*, p. 149.

40 Foucault, 'Questions on geography', p. 77.

41 Soja, *Postmodern Geographies*, p. 18.

42 See, for example, Kevin Hetherington, *The Badlands of Modernity: Heterotopia and Social Ordering* (London and New York, Routledge, 1997); Benjamin Gennochio, 'Discourse, discontinuity, difference: the question of "Other spaces"', in Sophie Watson and Katherine Gibson, eds, *Postmodern Cities and Spaces* (Oxford, Blackwell, 1995); Edward Soja, 'Heterotopologies: a remembrance of other spaces in citadel-LA' in *Postmodern Cities and Spaces*; and Georges Teyssot, 'Heterotopias and the history of spaces', *Architecture and Urbanism* 121 (1980), 79–100.

43 Foucault divides heterotopias into two broad types: those of crisis, and those of deviation, although this is another rather undeveloped point in the essay.

44 Another example is the garden, a heterotopia with 'very deep and seemingly superimposed meanings' (OS, p. 25). Foucault mentions Persian gardens, various areas within which symbolise different aspects of the world, such as sacred or forbidden spaces.

45 Hetherington, *Badlands*, p. viii.

46 Hetherington, *Badlands*, p. 7, stresses the processual nature of heterotopic ordering.

47 The now classic reading of this sort of interconnection is found in Edward Said, *Culture and Imperialism* (London, Vintage, 1994), pp. 20–35.

48 Foucault, *The Order of Things: An Archaeology of the Human Sciences*, trans. A. M. Sheridan (London, Routledge, 1970), p. xvii.

49 Foucault, *Order of Things*, p. xviii.
50 Foucault, *Order*, p. xviii.
51 The notion of heterotopia is perhaps also found in another 1966 essay by Foucault, on Blanchot, entitled 'Thought of the outside'. Foucault twice uses the term 'placeless places' to describe Blanchot's use of language. See *Essential Works of Foucault*, vol. 2: *Aesthetics, Method and Epistemology*, trans. Robert Hurley and others (London, Allen Lane–Penguin, 1998), p. 153.
52 Genocchio, 'Discourse, discontinuity, difference', p. 43.
53 See, for example, Shields, *Places on the Margin*.
54 Hetherington, *Badlands*, p. 49.
55 Michel de Certeau, *The Practice of Everyday Life*, trans. Steven Randall (Berkeley, University of California Press, 1984).
56 De Certeau's focus is upon the *parole* of everyday practices, not the social *langue* favoured by Saussure and structuralist linguistics.
57 Thomas Hardy, *Tess of the D'Urbervilles* (Harmondsworth, Penguin, 1978), pp. 48–9.
58 For a more extended critique of the cartographic claim to objectivity see J. B. Harley, 'Deconstructing the map', in John Agnew, David N. Livingstone and Alisdair Rogers, eds, *Human Geography: An Essential Anthology* (Oxford, Blackwell, 1996).
59 Arguably, this identification of place is a discourse that Hardy utilises not only in landscape description but also in the rendering of character: many passages in Hardy describe features of the human body and its clothing as if there are co-ordinates upon a map with a precise individual significance that is completely 'legible' to the author and reader. For one example see the opening page of *The Mayor of Casterbridge* (1886).
60 James Joyce, *Ulysses*, ed. Jeri Johnson (Oxford, Oxford University Press, 1993).
61 Harvey, *Postmodernity*, p. 240.
62 Harvey, *Postmodernity*, p. 240.
63 Marshall Berman, *All that Is Solid Melts into Air: The Experience of Modernity* (London and New York, Verso, 1983), pp. 15–16. For a critique of Berman see Perry Anderson, 'Modernity and revolution', *New Left Review* 144 (March–April 1984), 96–113. For an overview of this debate see Peter Osborne, 'Modernity is a qualitative, not a chronological, category', *New Left Review* 192 (March–April 1992), 65–84.
64 Harvey, *Postmodernity*, p. 99.
65 Harvey, *Postmodernity*, p. 257.
66 Harvey, *Postmodernity*, p. 264.
67 Harvey, *Postmodernity*, p. 264.
68 For an argument that the cultural significance of the experience of speed predates Harvey's sense of modernity see Jeffrey T. Schnapp 'Crash (speed as engine of individuation)', *Modernism/Modernity* 6: 1 (January 1999), 1–50.

69 Joseph Conrad, *Heart of Darkness*, ed. Robert Kimbrough (New York, Norton, 1988), p. 11.
70 Conrad, *Heart of Darkness*, pp. 10–11.
71 Harvey, *Postmodernity*, p. 265.
72 Harvey, *Postmodernity*, p. 270.
73 For a more positive reading of Pound's engagement with non-Western models of culture see Helen Carr, 'Imagism and Empire', in Howard J. Booth and Nigel Rigby, eds, *Modernism and Empire* (Manchester, Manchester University Press, 2000).
74 Harvey, *Postmodernity*, p. 272.
75 Harvey, *Postmodernity*, p. 273.
76 The classic account of this position is Georg Lukács, 'The ideology of modernism' in *Realism in Our Time: Literature and the Class Struggle*, trans. John and Necke Mander (New York, Harper & Row, 1971).
77 Harvey, *Postmodernity*, p. 273.
78 Virginia Woolf, 'Mr Bennett and Mrs Brown' in *Selected Essays*, vol. 1: *A Woman's Essays*, ed. Rachel Bowlby (London, Penguin, 1992), p. 70. For a cultural history of 1910 see Peter Stansky, *On or About December 1910: Early Bloomsbury and its Intimate World* (Cambridge, MA, Harvard University Press, 1997).

2 Through modernity: Forster's flux

The closing words of any novel always repay close attention, and E. M. Forster's conclusion to *Howards End* (1910) is no exception. The final words indicate the place and date of composition: 'Weybridge, 1908–1910'.[1] Rooted in a small southern town on the edge of the London metropolis, where the author lived from 1904 to 1925, Forster's closing words invite us to read his text as a specific exploration of the dynamic geography of modernity and of the place of the novelist in that world. The dialectic of space versus place, characterised by Harvey as one of the key features of modernism, infuses Forster's entire novel. For if it closes with a seemingly emphatic discourse of place, it is the connection of this figure with what de Certeau terms the 'actualization of space' that is the most interesting feature of the novel.

'Only connect', stated Forster in *Howards End*. But what exactly is to be connected, and more significantly, what is the manner of the connections exemplified in Forster's engagement with modernity? This chapter explores the diverse geographies depicted in the novel and interprets the key theme of connection in a manner that differs from the orthodox by stressing the idea of *spatial* connections. Forster's metaphor of connection – of conjoining the various binary terms, such as the prose and the passion, the spiritual and the material, and the Wilcoxes and the Schlegels, that the novel entertains – is related to a movement between a range of *topoi* and geographies. *Howards End* is a polytopic novel that illustrates Lefebvre's concept of the hypercomplex nature of social space. It is often interpreted as a 'condition of England' novel, a genre of fiction dating from the mid-nineteenth century, and revived from the 1890s to the late 1910s. The genre attempted to analyse, in fictional form, national problems about society.[2] The condition of England that Forster scrutinises seems structured by a dialogue between place and space, in de Certeau's terms, or between fixity and flux, to use Forster's terms. In the manuscript draft of the novel Forster noted that the house at Howards End does not represent 'a movement, for it rests on the earth'.[3] Although Forster seems instinctively drawn to an identification of place, *Howards End* cannot be viewed as a nostalgic desire to 'dwell', in the Heideggerean sense of place.[4]

Modernity, with its characteristics of movement, speed and the furious restructuring of spaces, cannot be avoided. As Daniel Schwarz notes of Forster's fictional engagement with the instabilities of the modern world: 'Flux is both an inevitable part of life with which we must come to terms and an enemy which we must combat.'[5] Sometimes in *Howards End*, however, it seems that flux is an enemy that must, in some fashion, be embraced: 'Only connect' may thus be read as a verb of movement across spaces, and not just as an exhortation to link two fixed places together.

Central among the range of spaces surveyed in Forster's novel is the city of London, being busily reconstructed due to extensive rebuilding work in the period prior to 1914. Surrounding the city is the 'creeping red rust' of the suburbs inhabited by Leonard Bast and his lower-middle-class clerkly compatriots, ensconced in the south and east of the city at a safe distance from the affluent west end inhabited by the Schlegel and Wilcox families. Forster is careful to spell out the social and class nuances of different regions of London. Leonard, for instance, is first impressed by the Schlegels only after seeing their card with its west end address (*HE*, p. 49). The suburbs also intrude on the country retreat of the Wilcoxes, the eponymous Howards End, and at several key points in the novel Forster eulogises the southern English landscape, particularly at Swanage. The English landscape, however, is part of a novelistic meditation upon the spaces of national identity, the *Englishness* of the Wilcoxes being juxtaposed against the Germanic Schlegels in a number of places. Another key setting in the novel is Oniton, a place on the English–Welsh border that again provides Forster with an opportunity to discuss the national characteristics of certain geographies as well as render in material form the notion of the border that must be crossed if any connection between two places or peoples is to be formed.[6] These national spaces are juxtaposed against repeated references to 'cosmopolitanism', Forster's synonym for modernity itself, where the upheavals of the contemporary world produces inhabitants of a cosmopolitan hue rather than citizens who belong in any specific place.

Many other spaces suffuse this novel, and at times it seems as if Forster can only sketch what these are rather than assemble them together in any meaningful arrangement. There are the brief references in the novel to imperial space and England's historical role at the turn of the century in building up its Empire by forcible and economic conquest. This is most noticeably present in the 'colonial spirit' (*HE*, p. 204) of the male members of the Wilcox family, who all have some association, through business or work, with overseas dominions. In comparison to these broader external spaces we have the inner domestic spaces of rooms and houses, most centrally in the symbolic role played by the house named Howards End.[7] Importantly, it is here that we witness a concern with the gendering of space, with Howards End possessing a significantly female set of meanings, from the first Mrs Wilcox's ownership to the way that the Schlegel sisters feel intuitively that the house is theirs.

Before discussing how Forster attempts to find connections between these various spaces and places, and how this is the key to understanding the novel's encounter with modernity, it is worth drawing attention to the spatial form of the text itself. Some critics have tended to see Forster's work before the First World War, and *Howards End* in particular, as representative of a strain of anti-modernist writing in English fiction, a form of fiction unable to take the plunge and dive into the maelstrom of modern experience in the manner of, say, Joyce or Richardson. As Langland notes, these interpretations of Forster imply that he has 'recourse to a nineteenth-century liberal humanism in resolving his novels', a strategy that drastically distances him from the mainstream of literary modernism.[8] However, critics such as Jameson, Langland, Meisel, and May have all sought to challenge this view of Forster, claiming of *Howards End* for example as Meisel does, that the novel 'seeks to catalogue and question the chief ideological assumptions that govern modernist speech' and that it is a 'meditation on modernism at large'.[9] It is in this spirit of a requestioning of received notions of literary modernism, and of Forster's place within that history, that I analyse the geographies of modernity within *Howards End*. This reveals a Forster whose encounter with the multiple spaces of modernity shows his work not to be pre-modernist, or nearly modernist, or even anti-modernist, but rather as engaged in producing an *attitude* towards modernity different from that exemplified by writers such as Woolf or Joyce.[10] This different attitude manifests itself primarily in the novel's attention to material and metaphorical spaces.

Richard Sennett, in *Flesh and Stone*, interprets *Howards End* as a key text for what it reveals about early twentieth-century attitudes in Western Europe to the city. Sennett interestingly suggests that Forster's novel gestures towards a 'modern meaning of place' in which the 'discontinuity' of modernity 'becomes a positive value'.[11] People are forced across social and sexual barriers (for instance, the child born to the middle-class Helen Schlegel, fathered by the poor clerk Leonard Bast, and who finally inherits the house), and the sense of place as fixity has to be connected with the fluidities of the modern world. One achievement of *Howards End* lies in the way the text represents the spaces of modernity in a deeply ambivalent fashion: aware of their presence to the extent that they cannot be ignored, but unable to find therein much to celebrate or embrace in fictional form. This ambivalence itself, I would argue, is a key part of modernism and should be recognised as such in any fresh understanding of modernist writing.

For Forster the city of modernity perceived in London in 1910 represents the development of a 'nomadic civilization' (*HE*, p. 256) or a 'civilization of luggage' (p. 154);[12] one of the most significant features of the novel is its attempt to understand this sense of the movements of modernity. This it does by trying to represent a modernity where 'all that is solid melts into air'.[13] Forster is obsessed with the notion of the nomadic basis of modernist identities, even though his attitude towards them is defiantly not one of simple celebration, as in certain forms of postmodernist thought. 'Flux' is a key term in *Howards End*: 'I hate this continual flux of London', comments Margaret Schlegel, for it is 'an epitome of us at our worst – eternal formlessness' (p. 184). Forster struggles to represent this key set of experiences, to discover a literary form to capture the fluxus of modern life, and to uncover a way to connect the modern writer to this central current of contemporary life.

Connection, then, is not a mere description of human relationships in the novel. We should take it at times in a brutal and literal fashion: this is a novel about conjoining different forms of space, and about the experience of moving between these spaces in the process of making connections. This, as I show below, is seen most clearly in the treatment of transport, particularly that of the motorcar. *Howards End* is a novel that yearns to connect together the modern metropolis, the new Edwardian suburbs, the pastoral landscape of the English countryside and the imperial domains upon which so much of the wealth of the European empires was based. In this ambition the novel maps the way for later modernist attempts to capture the motions of modernity; Forster's map is only a provisional guide, and shows the ambivalence felt about the new social spaces of modernity.

First, I discuss how Forster represents the spaces of the metropolis and its suburbs, before looking at how the novel conceives of national and border spaces. The chapter finally considers the most potent image of movement through modernity in the novel – the motorcar.

Flux, form and place

Forster's struggle to capture the experience of modernity is clearly seen in the depiction of the geography of London, particularly around the conceptual opposition of *space* and *place*. The urban historian Lewis Mumford, in his seminal *The City in History*, comments: 'Human life swings between two poles: movement and settlement.'[14] Forster's novel endorses that view, and often contrasts specific places with cities or areas that can only be known as shifting spaces. Whereas, for example, the southern seaside town of 'Swanage, though dull, was stable', London, in comparison, represents an uprooted existence: 'London only stimulates, it cannot sustain' (*HE*, p. 155).

However, *Howards End* is not an intrinsically anti-urban text, as Forster's rather crabby attitude sometimes suggests, but a complex response to modernity that keeps returning to city spaces in repeated attempts to connect them to wider themes in the novel. The city is in a process of flux, subject to the constant rebuilding of its material spaces; this flux produces the characteristically urban experience of disorientation and upheaval. Consider, for example, this early description of London:

> Over two years passed, and the Schlegel household continued to lead its life of cultured but not ignoble ease, still swimming gracefully on the gray tides of London. Concerts and plays swept past them, money had been spent and renewed, reputations won and lost, and the city herself, emblematic of their lives, rose and fell in a continual flux, while her shallows washed more widely against the hills of Surrey and over the fields of Hertfordshire. This famous building had arisen, that was doomed. Today Whitehall had been transformed; it would be the turn of Regent Street tomorrow. And month by month the roads smelt more strongly of petrol, and were more difficult to cross, and human beings heard each other speak with greater difficulty, breathed less of the air, and saw less of the sky. Nature withdrew. (*HE*, p. 115)

Clearly the description is coloured by a dislike of the metropolis, signalled in the use of the extended metaphor of the sea and the movement of water. The city possesses 'gray tides' on which the Schlegel family move, but the trope alters somewhat when the family merge into the city. London is the emblem of their lives, and its 'continual flux' of urban expansion into the suburbs ('her shallows') and rebuilding gradually loses its oceanic quality as Forster sees only the negativity of urban life. Traffic pollutes and blocks the roads, affecting the city-dweller in a quite physical way: people cannot hear each other, and thus communication is thwarted.

The image of the city as 'emblem' is problematic, and Forster is unable to sustain the comparison with the Schlegel family. He cannot, once nature withdraws, conceive of a new set of tropes with which to understand the oscillations of modern life. The section following this description heightens the difficulty of registering the spaces of the city. Forster's narrator notes that it is not fashionable to decry London and that new literature will 'probably ignore the country and seek inspiration from the towns' (p. 115). So Forster then grapples to understand how London may be integrated into literature:

> Certainly London fascinates. One visualizes it as a tract of quivering gray, intelligent without purpose, and excitable without love; as a spirit that has altered before it can be chronicled; as a heart that certainly beats, but with no pulsation of humanity. It lies beyond everything: Nature, with all her cruelty, comes nearer to us than do these crowds of men. A friend explains himself, the earth is explicable – from her we came, and we must return to her. But who can explain Westminster Bridge Road or Liverpool Street in the morning – the city inhaling – or the same thoroughfares in the evening – the city exhaling her exhausted air? We reach in desperation beyond the fog, beyond the very stars, the voids of the universe are ransacked to justify the monster, and stamped with a human face. (p. 116)

Here Forster gropes towards an understanding of city space: the metaphors of nature are gone, but he is unable to articulate a discourse to replace them. London is defined paradoxically: it lacks colour, purpose, love and humanity, and although it does possess intelligence, excitement and pulsation, these qualities are drained of profundity. Significantly, the city has undergone changes, but these have not been properly 'chronicled'. This mournful note is perhaps a self-directed comment by the author, aware of a duty to 'chronicle' or narrate the city, but unable to keep up to speed with its ceaseless reshaping. Somehow the city is neither positive nor negative – it exists 'beyond everything', unable to enter fully into representation. Though the city surrounds us, it is nature that, metaphorically, is 'nearer to us': the metaphoric space of comprehension has deserted the material spaces of the city. We can understand the metaphor of life being a journey from and towards death, but the actual journeys of commuters at Liverpool Street railway station remain inscrutable. Forster refuses to let the image escape, however, and tries to anthropomorphise the city as a monster breathing in these commuters. But this attempt is only a sign of 'desperation', moving yet further away ('beyond the very stars') from the space one is trying to understand. Forster's final image of London as 'religion's opportunity' indicates that which is completely beyond, both spatially and conceptually. In having recourse to religion Forster demonstrates that he is unable, in the words of an earlier critic, Matthew Arnold, that are repeatedly quoted in the novel, 'to see life steadily and see it whole'.

The search for a settled discourse with which to view the city as a whole fails, as Forster acknowledges: 'The Londoner seldom understands his city until it sweeps him, too, away from his moorings' (p. 116). The instabilities of the modern city are the key to understanding London, but immersion within them is problematical. Margaret learns that the lease on the Schlegels' home in Chelsea is to expire and that the owner wants to rebuild the property as flats.[15] This production of space for economic reasons disorientates Margaret at the same time as it lays bare the true fluidity of the modern city: 'In the streets of the city she noted for the first time the architecture of hurry, and heard the language of hurry on the mouths of its inhabitants – clipped words, formless sentences' (p. 116). Movement – the hurried rebuilding of the city – intrudes even upon the language of modern life; but yet again Forster cannot reproduce such a discourse, only catalogue its existence.

Here we detect Forster's ambivalent desire to represent the spaces of modernity in a language appropriate to them. Can the novelist find a form in which to depict a formless world? Form stands as the binary opposite to flux, and the novel is aware of a need for a modernist form able to register this incessant movement. Margaret receives a troubling letter from Helen, who has fled abroad after becoming pregnant by Leonard Bast, and her feelings for her sister are interfused with her view of the city. She saw London as it really is, 'a caricature of infinity' (p. 275). Saddened by this revelation, Margaret, while en route to her husband Henry's office to get advice on how to handle Helen, goes into St Paul's Cathedral 'whose dome stands out of the welter so bravely, as if preaching the gospel of form' (p. 275). Wren's classical cathedral represents an earlier and more stable vision of the city. But this attempt to juxtapose built form against infinity, or flux, fails, and the inner space of the cathedral 'is as its surrounding ... and points us back to London'. The exterior hubbub of the city intrudes upon the secluded space of the cathedral, illustrating Lefebvre's notion of the 'ambiguous continuity' between seemingly disjunct spaces. Helen will not be found inside the cathedral, and her disruption of the family's life is thus associated with the formlessness of the city. The narrative events of the novel are thus embedded within the material spaces represented in the text. For Margaret, 'Helen seemed one with the grimy trees and the traffic and the slowly flowering slabs of mud' (p. 275).

If Forster cannot imagine a form in which to capture this urban flux, he is at least aware that this quest cannot be perpetually ignored: even the classicism of St Paul's points back to the infinity of London. We might take this as a comment on the textual space of *Howards End*: unable to retreat into an older narrative form, but as yet not fully cognisant of a new form. Forster's temporary withdrawal, after *Howards End*, from writing novels can be seen in this light, as an intuition into the gap between the flux of modernity and the form of the modern novel, and an awareness that he was, as yet, unable to bridge this gap, or, we might say, unable to connect the two spaces – of modernity and the novel.[16]

Howards End does, however, contain many interesting discussions of place and its relation to space. In chapter 15 an important chance meeting occurs between Henry Wilcox and Margaret and Helen, while walking along the River Thames at Chelsea Embankment. Margaret and Helen request Henry's advice regarding their new friend Leonard Bast, and are told that Bast's employer's company, the Porphyrian Fire Insurance Company, will very soon be bankrupt. The sisters convey this advice to Leonard, but unfortunately the tip turns out to be incorrect, Leonard loses his job at the new company he has joined, and the subsequent revelation of Henry's affair with Leonard's wife, Jacky, unfolds as a consequence. The setting on Chelsea Embankment is important, for it is one of the few unambiguously positive images of the city in the novel. This, the author notes, is 'open space used rightly' and one which produces a 'note of dignity that is rare in English cities' (p. 135). The affirmative qualities of this space, created in the 1870s, relate to its proximity to the natural image of the river and because it recalls other European cities, especially Germany. The description utilises a discourse of place, in de Certeau's terms, for the city is imagined here as a theatrical spectacle to be visually consumed: 'the city ... seemed to be a vast theatre, an opera-house in which some endless trilogy was performing, and they themselves [Margaret and Helen] a pair of satisfied subscribers' (p. 135). Unlike the inexplicable London of morning commuters at Liverpool Street, Chelsea Embankment is a reified visual object to be watched with pleasure.

Place is an important topic of conversation between the Schlegel sisters and Henry, not only because of a need to find a *place* of work for Leonard Bast, but also because the Schlegels are to leave Wickham *Place* (the name itself indicating a rooted location) and, we learn here, because the Wilcoxes have recently vacated Howards End. Margaret and Helen have just left a discussion circle of liberal friends where the topic of conversation had been how to dispose of unearned income so as best to help disadvantaged people like Bast. In an intriguing image the sisters link money as the warp of the world with a sense of place as the woof of the world (p. 136), connecting the economic themes of the novel – including the plight of Bast – with the representation of space.[17] Margaret ponders the significance of place when she notes: 'It is sad to suppose that places may ever be more important than people' (p. 136). She then elaborates on this theme to Helen: 'I believe we shall come to care about people less and less ... The more people one knows the easier it becomes to replace them. It's one of the curses of London. I quite expect to end my life caring most for a place' (pp. 136–7). Bast, of course, is one of those replaceable people and his fate is explained as another aspect of the flux of the city itself. Margaret's claim also anticipates the ending of the novel, where she does indeed care more for a place – Howards End – than a person such as Bast, or even her husband.

Ironically, it is with these words that Forster reintroduces into the narrative her future husband, Mr Henry Wilcox, who also cares more for places than he does for people. Henry's care for places, however, is connected to his perception of their economic worth, rather than their cultural or personal value. This distinction is made by reference to the immediate surroundings of this encounter by the Thames, showing how social space feeds the metaphorical needs of Forster's text. For Henry economic success means that the 'world seemed in his grasp as he listened to the river Thames which still flowed inland from the sea' (p. 137). In Conrad's *Heart of Darkness* the Thames sprawled outwards to Africa; for Henry the river flows 'inland', symbolising the wealth it brings him from his imperialist work in African rubber. Henry's attitude to the river is thus one of financial practicality: 'So wonderful to the girls, it held no mysteries for him' for he had 'helped to shorten its long tidal trough by taking shares in the lock at Teddington, and, if he and the other capitalists thought good, some day it could be shortened again' (p. 137). This again is part of the imagery of urban flux, with the natural tidal movement of the Thames being subdued by the requirements of capitalist trade, the woof of place being ruled by the warp of money. Forster's discourse of place – the eulogy to Chelsea Embankment and the comments by the Schlegels – is always balanced by a Lefebvrean sense of the production of social space.

Henry's actions, which are those of a man of business, oppose the rootedness in place that the Schlegels represent. The Wilcoxes, we are later told, have no part 'in any place' (p. 246), but, as one character comments, 'They keep a place going' (p. 268) by financing the restructuring of spaces, urban and otherwise. Rather than Margaret's connective affinity for Howards End, Henry's view of the question of who is to inhabit the house is 'connected with ... the rights of property' (p. 317). When she learns that Ducie Street, where the Wilcoxes had lived, is to be demolished to be rebuilt as mews, Margaret expresses regret. Henry's attitude is more phlegmatic: 'Shows things are moving. Good for trade' (p. 184). After the incident on the Chelsea Embankment, Henry leaves the sisters watching the ebbing of the tide with a cheery comment – 'Everyone moving!' – on learning that both families are moving house. Margaret turns this into another metaphor for modern life: 'Everyone moving. Is it worth while attempting the past when there is this continual flux even in the hearts of men?' (p. 143) 'Attempting' is a typically obtuse Forsterian term, but here indicates how the spatial experience of modernity triumphs over any temporal sense of tradition. Henry's identification with these movements later leads an exasperated Margaret to ask, upon learning that they are not to live at Oniton Grange: 'Don't you believe in having a permanent home, Henry?' (p. 255).

People move house repeatedly in *Howards End*, emphasising how modernity disrupts a stable sense of place. Protesting at such upheavals, the first Mrs Wilcox queries whether modern civilisation can be right if 'people mayn't die in the room where they were born' (p. 93). But we also learn that the city itself is moving, being rebuilt by speculators approved of by Henry. Both Ducie Street and Wickham Place are rebuilt as flats to generate greater income. Even more indicative of the flux of the city, however, is the suburban housing of Leonard and Jacky Bast. The rise of the suburbs, and the cultural ramifications thereof, were a dominant preoccupation in Edwardian and Georgian England, and some critics have argued that modernism owes some of its aggressively metropolitan *brio* to a crusade against the cultural deadening of suburban life.[18] For writers like Ezra Pound and Wyndham Lewis the suburbs represented the antithesis of all that avant-garde modernism promoted; the suburbs meant an essentially feminised culture of the mass market, with low-brow tastes, a petty morality and an overarching narrowness and uniformity of vision.[19]

Forster's novel shares this negative judgement, following the lead given by social commentators like his friend C. F. G. Masterman in *The Condition of England* (1909) and T. W. H. Crosland's *The Suburbans* (1905).[20] The Basts' first home is in south-east London, probably Brixton or Stockwell, and certainly a far different setting from that of the Schlegels' home on the north side of the Thames. A map of the London locations of the novel, such as those drawn by Moretti in his *Atlas of the European Novel*, would reveal how Leonard's life is geographically, as well as socially, distinct from that of the Schlegels or the Wilcoxes. All of the action involving these two families takes place north of the Thames, whether in the somewhat artistic area of Chelsea, where the Schlegels live, or amid the more aristocratic wealth of the Wilcoxes' Belgravia home in

Ducie Street. Bast crosses the river to the north; but the two respectable families never cross over to Leonard's side of the river.

The Basts' first home, in 'Camelia Road', is even further removed, geographically and symbolically, from Howards End, located north of London in the village of Hilton, Hertfordshire, and based on Stevenage in real life. Later the Basts move slightly eastwards to Tulse Hill (p. 130), and it is in such areas that contemporary commentators habitually placed suburban clerks. For Masterman, they are to be found 'covering the hills along the northern and southern boundaries of the city' in such areas as Clapham and Herne Hill.[21] Crosland also identifies these southern and south-eastern areas as essentially suburban, describing Clapham as 'the capital of suburbia' and Tooting as 'a whirling wilderness of villadom'.[22]

Forster's description of the exterior of Bast's suburban home continues the theme of urban flux, albeit in a differently classed location:

> A block of flats, constructed with extreme cheapness, towered on either hand. Further down the road two more blocks were being built, and beyond these an old house was being demolished to accommodate another pair. It was the kind of scene that may be observed all over London, whatever the locality – bricks and mortar rising and falling with the restlessness of the water in a fountain, as the city receives more and more men upon her soil ... Plans were out for the erection of flats in Magnolia Road also. And again a few years, and all the flats in either road might be pulled down, and new buildings, of a vastness as present unimaginable, might arise where they had fallen. (p. 59)

Again Forster's image of modernity is figured in terms of fluidity, here the water in the fountain. The message is the same as that of the earlier comparison of London as a whole to a sea: this is space being incessantly recomposed in such a way that settlement, or a Heideggerian attachment to place, is near impossible.

In another sense suburbia itself represents a slightly different symptom of flux, for, as Flint comments, '[s]o rapid was the growth of London's suburbia that its precise location had to be continually redefined'.[23] Suburbia did not just mean the rebuilding of city housing, but also the city moving outwards to devour the countryside. As early as 1901 Masterman noted this lateral 'spreading' of the suburbs: 'North, East, South, and West the aggregation is silently pushing outwards like some gigantic plasmodium: spreading slimy arms over the surrounding fields, heavily dragging after them the ruin of its desolation.'[24] Forster also comments upon this phenomenon, most notably in one of the closing images of the novel, when a mechanical trope replaces Masterman's organic slime. From the rural vantage point of Howards End, across 'eight or nine meadows', there is a 'red rust' representing London 'creeping' (p. 329) over country areas, both north and south, such as Surrey, Hampshire and Hertfordshire. This doleful comment echoes one of Margaret's first impressions of the village near Howards End, that it had not yet decided whether to develop into 'England or Suburbia' (pp. 29–30). In this binary the geographical experience of the suburbs is excluded from Forster's definition of England; England is signified by a rural space, but one under threat from the spaces of city and suburb.

Although Forster wishes to show how the rush of modernity is equally experienced by classes other than those of the independently wealthy Schlegels or business tycoons like the Wilcoxes, he is somewhat unable to connect the novel's textual space to Bast's suburban class. At a narrative level this is felt in the rather forced and abrupt handling of key events in the novel: the brief liaison between Leonard and Helen; the revelation of the affair between Henry and Jacky Bast; and Helen's subsequent child by Leonard. These events literally embody connection between the upper classes and the culture of the clerk.[25] But at this point the reader's incredulity in the plot matches the spatial distance between the Basts and everyone else in the novel. It is significant that in a novel where *topoi* are so rich in connotation Forster gives only one vivid description, in chapter 6, of where Leonard Bast lives. When the Basts move to Tulse Hill there is no depiction of their new residence. The Basts are always being dragged out of their space into the geographies of other classes, as when Leonard first meets the Schlegels after the concert and is taken to Wickham Place to retrieve his umbrella. Helen also takes Leonard and Jacky to Oniton, in Shropshire, and Leonard is forced to travel to Howards End – and his own demise – to find out about Helen.

Forster, it seems, cannot move into the Basts' space, and is forced to pull them out of it to propel the novel's plot forward. When they do leave the suburbs the Basts only signify to the two main families a sense of 'the abyss', the term used from the 1890s by social commentators such as Masterman to refer to working-class zones of London, mainly gathered in the east of the city.[26] Early on we learn that Leonard 'was not in the abyss, but he could see it, and at times people whom he knew had dropped in it' (p. 58). To the other families the abyss is recognisable by smell. When Leonard calls on the Schlegels in order to clear his name with his wife, Margaret is 'distressed again by odours from the abyss' (p. 124), an image re-used when the Basts turn up at Oniton. The abyss, in proximity to the suburbs, is a kind of blank space in the novel, indicating somewhere that

Forster cannot really represent as a place, only as a kind of malodorous unpleasantness.[27]

If the suburbs and the attendant abyss are a kind of external limit to Forster's spatial connection, the interior of the Basts' home is used to signify further the limitations of suburban culture. As David Trotter has argued of interiors in the Edwardian novel: 'Rooms express the people who occupy them, and both are symptomatic of the condition of England.'[28] Bachelard's 'intimate space' of the room is here suffused with meanings drawn from exterior social locations. The Basts' residence is a three-roomed furnished basement flat (p. 61); a 1908 survey estimated that in London nearly 39 per cent of people lived in three-room accommodation, compared with the national average of only 18 per cent.[29] Although only two people, and not a family, live in this space, Forster employs the smallness of the space to render the narrowness of Leonard Bast's view of the world. The sitting-room contains a 'cosy corner', with the whole of one wall being 'occupied by the window' (p. 60); the overall effect of the flat is 'dark as well as stuffy', and Leonard daydreams of the Schlegels' home with its 'rich staircase' and 'some ample room' (pp. 62, 67). Cruelly, Forster contrasts the interior architecture of the Basts' home with the splendour of Venice, by having Leonard read Ruskin's *Stones of Venice*, showing not just his aspirations towards art and culture but the *physical* distance between this house and those buildings described by Ruskin.

The limits of Leonard's flat are the limits of his world, as Forster makes clear when he visits the Schlegels later and tells them of his walking escapade outside the city in the Surrey countryside: 'What's the good ... of living in a room for ever? There one goes on day after day, same old game, same up and down to town, until you forget there is any other game. You ought to see once in a way what's going on outside, if it's only nothing particular' (p. 127). The uniformity of suburban life is emphasised by the image of living in only a single room, dreaming of an outside. Leonard's escape, however, is illusory, and Forster transfers the confined interior space of his room into an image for Leonard's mental space: he has a 'cramped little mind', and in this mind 'he had ruled off a few corners for Romance' (pp. 127, 129), assigning the Schlegel sisters to this particular spot in his psychic life. In a later incident Leonard's very body seems imbued with the quality of this room: 'doors seemed to be opening and shutting inside his body' (p. 312). Bast's material environment, then, becomes the metaphorical space through which Forster designs his interior life, subtly blurring exterior and interior spaces. Leonard's death – crushed by a bookshelf inside the house at Howards End – epitomises his final confinement, pinned down by the cultural capital he dearly dreamed of, but which the narrative denies him.

Spaces of the nation

Although *Howards End* is haunted by images of flux and loss of relationship to place, it is revealing how Forster contrasts the city with images of rural spaces in which national identity is explored. If the suburbs and the abyss seemed somehow detached from the geography of Englishness, this is certainly not the case when Forster discusses the landscape of the Purbeck hills in Dorset or Oniton in Shropshire. The representation of these regions has a dual function: first, as a resistance to the flux of modernity symbolised by the city; second, as a synecdoche for the national space of England itself. However, neither of these aspects can avoid the movements of modernity, for both concern not just enclosed spaces but the borderlands around them. For the city–country couplet this, as shown above, meant a consideration of the 'red rust' of suburbia. The national space of England is similarly troubled by the image of the border, most specifically in the meditation on Saxon and Celt prompted by the boundary between England and Wales at Oniton. To that image of the border we should add the connections Forster examines between England and its imperial domains, figured in Henry's work in Africa.

There are only three major locations outside London represented in *Howards End*: the house at Howards End, in Hertfordshire north of London; Swanage and the Purbeck hills on the central south coast of England; and Oniton, a small town on the border between England and Wales.[30] Partly, this circumscribed geography indicates the limitations of Forster's 'condition of England' approach, 'England' presumably finishing somewhere north of Shrewsbury.[31] But the three locations are also significant for their position in the novel's spatial form: major narrative crises happen in all three places. At Howards End Helen is reunited with Margaret and Leonard is killed; at Swanage Margaret decides to marry Henry, and precipitates a temporary break from her sister; and at Oniton, the marriage of Evie and Charles is disrupted by Helen bringing the Basts to the celebration, an event which reveals Henry's past affair with Jacky Bast, and which also leads to Helen's fateful liaison with Leonard. This link between non-metropolitan places and narrative changes suggests that only when exterior to the flux of the city could Forster handle major moments of plotting. It is as if the slightly more settled spaces of provincial England can serve more smoothly as the backdrop for narrative surprises. They are also moments when Forster connects the major emotional crises in the inner life of his characters to his own sense of the problematical status of national identity: the textual space of the novel thus linking these two modes of crisis.

At the start of chapter 14, when Margaret retreats to Swanage to consider Henry's marriage proposal, the narrative commences with a seemingly marginal passage on the geography of the Purbeck hills: 'If one wanted to show a foreigner England, perhaps the wisest course would be to take him to the final section of the Purbeck hills' (p. 170).[32] There follows a eulogy to the Dorset countryside, to Salisbury Plain, the Isle of Wight, and 'all the glorious downs of central England'. Forster is, however, keen to represent something more than a merely pastoral version of Englishness. The ports of Portsmouth and Southampton, 'hostess to the nations', are included and, importantly, these spaces are linked back to the metropolis: 'Nor is suburbia absent. Bournemouth's ignoble coast cowers to the right, heralding the pine trees that mean, for all their beauty, red houses, and the Stock Exchange, and extend to the gates of London itself. So tremendous is the City's trail! (p. 170). The hold of London upon the outlying regions was also noted by the eminent geographer Halford Mackinder in his influential book *Britain and the British Seas* (1902): 'In a manner all south-eastern England is a single urban community ... The metropolis in its largest meaning includes all the counties for whose inhabitants London is "Town", whose men do habitual business there, whose women buy there.'[33] Forster's description is thus another example of the strongly geographical sense of *connection* that informs the text. The marriage proposal of the previous passage, between the more rural Margaret Wilcox and the resoundingly urban Henry, is thus depicted spatially by Forster's link between the Purbeck hills and the metropolis. If the national space of Englishness is to be comprehended, then, like the union of Henry and Margaret, seemingly different geographies must be conjoined.

Such a moment is seen in the final sentence of this opening paragraph, and shows Forster trying to join a particular place (Swanage) to a wider sense of social space (England):

> How many villages appear in this view! How many castles! How many churches, vanquished or triumphant! How many ships, railways and roads! What incredible variety of men working beneath that lucent sky to what final end! The reason fails, like a wave on the Swanage beach; the imagination swells, spread and deepens, until it becomes geographic and encircles England. (pp. 170–1)

It is significant that this picture is able to make connections, shifting from the rural (the villages and churches) to signs of technology and industry (ships, roads, rail). Naive ruralism this is not, but an image of a set of spaces, unlike London, that can be fully grasped, even if this is achieved not by reason but by a geographical imagination. Perhaps the sense of Englishness summoned up here is achievable precisely because the furious traffic of urban life is absent and, importantly, the narrating subject is detached from the scene below. In de Certeau's terms this is a discourse identifying a place, verging on the cartographic. From the Purbeck hills Forster can geographically 'encircle' England; from London he could only note the inconceivable metropolis as an unmappable set of spaces.

All through this chapter Forster associates the exterior spaces of the landscape with the interior life of his characters. Asked by her sister whether she loves Henry, Margaret finds 'her eyes shifting over the view, as if this country or that could reveal the secret of her own heart' (p. 176). This mode of spatial connection is also much in evidence in the crucial chapters set in Oniton, the key image being that of the border. Earlier, when Margaret first feels affection for Henry, across the literal space of their adjacent London dwellings, Forster notes: 'Her thought drew being from the obscure borderland' (p. 71). This border is between the individuals, and also between the national boundaries of England (Wilcoxes) and Germany (Schlegels) symbolically established in the novel. But the border also operates in a number of other important ways: Oniton, based on Clun in the Shropshire Marches, is a place where the secure Englishness of Dorset is challenged by the presence of Wales. Oniton was a market town which 'for ages' had 'guarded our marches against the Celt' (p. 208).[34] The Celtic fringe, in Irish rather than Welsh guise, was a dominant political issue in 1910, with the question of home rule for Ireland very much engaging public attention.[35] Margaret is fascinated by the 'romantic tension' (p. 216) of Oniton's border status, and she imagines settling there and converting Henry to a rural life. Margaret dreams of the 'power of Home' and of creating 'new sanctities among these hills' (p. 220). This identification with place is, however, shattered by the revelation, while at Oniton, of Henry's old affair with Jacky Bast. The uncertain space of the geographical border disrupts Margaret's cosy image of a place called home. As the humble Basts are brought into the aristocratic space of Oniton Castle, Margaret's inner life merges with the geographical history around her:

> Oniton, like herself, was imperfect ... It, too, had suffered in the border warfare between the Anglo-Saxon and the Celt, between things as they are and as they ought to be. Once more the west was retreating, once again the orderly stars were dotting the eastern sky. There is certainly no rest for us on the earth. (p. 229)

The Celtic west is unruly in comparison to the 'orderly' east, and Oniton proves not to be a place of 'rest'. The knowledge of Henry's affair resumes the 'border warfare' between Wilcoxes and Schlegels that the marriage was intended to conclude. Social space intrudes in a troublesome fashion into the personal space of relationships, and Margaret and Henry leave Oniton never to visit it again.

Commenting on how space is correlated to plot in nineteenth-century novels, Moretti notes that 'the crossing of a spatial border is usually the decisive event of the narrative structure'.[36] This is not quite the case in *Howards End* (which perhaps helps mark its distance from the nineteenth-century realist tradition) because the border with Wales is never really crossed. The incidents at Oniton are decisive, but they point once again to the problematical image of *connection*. England and Wales are not crossed or connected, but the Wilcoxes and Schlegels are finally conjoined with the Basts in the shape of Helen's child.

One way to interpret this moment is to see it as Forster's attempt to connect the various factions within English society by pushing them to the extreme geographical edge of England. This is no fixed place – like the landscape of Dorset – but a social space that brings with it a conflictual geopolitical history. Inner emotions and external loci coalesce in the textual space of the novel, and it is the *movement* across such spaces that becomes the defining motif of the novel. As Margaret had noted, there is to be 'no rest for us on the earth'. In the final section of this chapter I analyse a dominant motif of this movement through modernity in *Howards End* – the motorcar.

The motorcar

In a revealing essay on writing, Englishness and the cultural resonances of different forms of modern transport, John Lucas has drawn attention to how *Howards End* employs the motorcar as a 'symbol of intrusive, unsettling power' that disrupts the 'sentimental ruralism' Lucas detects in the novel.[37] However, the motorcar is not just a vehicle for symbolic meanings, as Lucas shows, but also has a powerful influence upon the narrative form of the novel itself. Likewise its role in the text is to represent more than just a nostalgic pastoral. In Forster's novel the speeding motion of the motorcar offers a significant image for comprehending the connections between the new spaces and spatial experiences of modernity. By focusing upon the motorcar's role in *Howards End* we can further situate the novel within a spatial history of modernity.

Throughout the novel the motoring exploits of the Wilcox family function as a profound symbol of technological modernisation and its attendant experience of modernity as *nomadic*. Forster's novel depicts a world where the invention of the internal combustion engine, at the end of the nineteenth century, resulted in far-reaching changes in the human experience of the basic categories of time and space. As one historian of the motorcar concludes: 'the motorization revolution ... is one of the major hallmarks of modernity, transforming social life [and] the economic system'.[38] In 1910, the year *Howards End* was published, Henry Ford set up his first European car plant in Manchester.[39]

Motoring in Britain in the period before the First World War signified a deep ambivalence over the course of the disturbances wrought by modernity.[40] At one level the expansion onto the roads represented an individual freedom (for the few wealthy enough to afford cars) previously little known to those accustomed to travelling by carriage or train. One guide to motoring in 1914 praised the fact that motor travel avoided the restrictions to personal liberty of 'the "time-table" journey' of the railway.[41] Another supporter wrote in 1909 that 'the day of the complete triumph of the motor is at hand' and that in comparison to horse-drawn travel 'the distance traversed [by car] is unlimited'.[42] Here the personal liberties proffered by motoring are clearly linked to modifications in the perception of space and time, with distant locations now being made much more speedily available. Motoring also represented speed and power, qualities redolent of modernity and change itself, seen in this description of a first car journey by a writer in 1908: 'Most marvellous of all perhaps were the grand obedience which this instrument gave him ... the new power it placed at his disposal, the new sensations which it begot, and the new situations which it created and opened out for him'.[43]

Such celebrations presage, prosaically, the famous praise for the car found in the Italian avant-garde movement of Futurism. Marinetti, founder of Futurism, first visited London in 1910, and his proselytising for the movement made a great impact on English artistic groupings.[44] The Futurist founding manifesto of 1909, composed by Marinetti, contains a lengthy description of racing a car through city streets, only to end up crashed, in a ditch, to avoid hitting two cyclists, hapless symbols of an archaic form of life standing in the way of the modern automobile. The manifesto continues by endorsing one of the main features signalling the modernity of the automobile – its speed: 'We affirm that the world's magnificence has been enriched by a new beauty: the beauty of speed ... We want to hymn the man at the wheel, who hurls the lance of his spirit across the Earth, along the circle of its orbit.'[45] Speed here is closely linked to a geographical sense of the car's vauntless possibilities in criss-crossing the globe, a point found repeatedly in contemporaneous accounts of motoring.

This awareness of spatial conquest by the car is associated with two other prominent geographical discourses in this period: tourism and imperialism. R. P. Hearne noted that 'the joy of the motorist was as that of a conqueror' and announced with satisfaction that 'the heart of Africa has been penetrated' by the car.[46] Edith Wharton, in *A Motor-Flight Through France* (1908), celebrated how the 'motor-car has restored the romance of travel'.[47] Berriman's 1914 *An Introduction to the Car and the Art of Driving It* includes a number of 'key-maps' of drives around London and elsewhere and devotes a chapter to 'touring' in which he argues: 'The motorist, unlike the tourist by rail, finds the best part of his holiday in the journey itself. The best of scenery is never so realistic or picturesque when seen through the framed glass of a railway carriage window as it is from the seat of an open touring car.'[48] This sense of the tourist gaze indicates an important element of modernity also found in *Howards End*'s emphasis on a 'nomadic civilisation', for, as a later commentator suggested, the motorcar brought with it an increased sense of 'through traffic'.[49] People now increasingly passed through places, perhaps glancing briefly at the sights, but not entering into any more permanent relationship with these towns or villages than that of the speeding voyeur. It is that fresh experience of space against which Margaret Schlegel rails towards the conclusion of *Howards End*, when she views the growth of suburbs around London, themselves partly created by the new access to the city offered by the car: 'This craze for motion has only set in during the last hundred years. It may be followed by a civilization that won't be a movement, because it will rest on the earth' (p. 329).

Margaret's criticism is perhaps more refined than those of other contemporaries, but it does share an unease with the consequences, both social and psychic, of the growth in car ownership. Between 1905 and 1910 car and commercial vehicle ownership increased from around 32,000 to a phenomenal 143,000, putting Britain far ahead of all other countries in the world, except for the USA, in the number of vehicles on the roads.[50] This period saw the motorcar as the subject of fierce debates about the disruptions it threatened to established patterns of life, fears that form the backdrop to Forster's negative image of the car and provide a contrast to claims made by those who eulogised the automobile. Here again we see that the focus is often upon the changes in social space represented by the car. If, for some, the motorcar was the 'herald of modernity', for others, motoring and motorists were a menace, 'cads on casters' as *Punch* termed them in 1896,[51] drawing attention to the wealth required to own and maintain a car. Car ownership fed Edwardian anxieties about the changing composition of the English class structure in the period, threatening old alliances and unsettling many liberal middle-class commentators.[52]

Forster's criticism of the car consists of two interrelated elements. First, it is directed at the financial privilege associated with car ownership, the clear identification of the car with the financial magnates, the Wilcoxes, emphasising that association. This view was also found in the writings of prominent social commentators such as C. F. G. Masterman.[53] In *The Condition of England* Masterman attacks the ostentatious 'waste' of Edwardian wealth, seeing the motorcar as one such emblem: 'The action of a section of the motoring classes ... in their annexation of the highways and their indifference to the common traditions, stands almost alone as an example of wealth's intolerable arrogances'.[54] The Automobile Association in 1904 suggested that the average cost of a car was £300, with a yearly expenditure of around £500.[55] In *Howards End* we learn that Margaret Schlegel's yearly income from investments is £600 (p. 72), a figure that puts her comfortably above Leonard Bast, but much below the car-owning Wilcoxes.

Second, Forster's portrayal of the 'throbbing, stinking car' (p. 36) stems from his perception that the car is part of a 'new civilization' (p. 10) that he found antipathetic at worst, bewildering at best. In this sense Forster's attitude recalls that of another eminent Edwardian, Kenneth Grahame, who uses the motorcar as an image that disrupts the golden world of *Wind in the Willows* (1908). Behind the wheel, Toad becomes 'the terror, the traffic-queller, before whom all must give way or be smitten into nothingness and everlasting night'.[56] Throughout the early years of the twentieth century the 'terror' of the car signified in material form the very course of modernity itself. One dimension of the anxiety attached to these changes is apparent in the discourses around the accidents, the speed and the dust of the motorcar.

Reckless driving caused a number of infamous accidents, such as the death of a child brought about in 1905 by the son of Alfred Harmsworth, the publisher.[57] In *Howards End* there are two significant car accidents, both used to symbolise forthcoming human tragedies. Immediately before the death of Mrs Wilcox, in chapter 11, Evie and Henry cut short their motoring holiday in Yorkshire due to a 'motor smash' (p. 96) with a horse and cart. Later, when Margaret travels to Oniton for Evie's wedding the car in which she is travelling runs over and kills a cat. Despite Margaret's protestations that they should stop in order to talk to the cat's owner, Charles refuses and Margaret jumps out of the moving car, cutting her hand (p. 212). At a simple level Margaret's behaviour displays her faith in intuitive human emotions against the coolly rational Wilcoxes. Margaret feels that, although 'she had disgraced herself', the girl whose cat had died had 'lived more deeply' than the Wilcoxes and others in the car: 'she felt their whole journey from London had been unreal. They had no part with the earth and its emotions. They were dust, and a stink, and cosmopolitan chatter' (p. 213). The accident thus indicates what is lost in the espousal of the values symbolised by the car. It also anticipates the coming revelations at Oniton.[58] The alienation Margaret feels as a result of the motor drive is also an estrangement from the 'cosmopolitan', a shadowy figure in the novel that indicates, yet again, the absence of an autochthonous sense of place.

The speed of cars, and the fact that the low speed limits were regularly broken, also brought concern for the Edwardians. In 1896 the Red Flag Act repealed the British speed limit of 2 mph and set a new limit of 12 mph.[59] The speed limit was increased again in 1904 to 20 mph, with local councils able to impose a 10 mph limit if they wished.[60] This new speed limit was widely ignored, since cars could already manage speeds of over 120 mph by 1908,[61] and the police in certain counties of England pursued speeders by setting numerous traps for them. In *Howards End* Mrs Wilcox refers to the police traps ('nearly as bad as in Surrey') as one reason for her husband cutting short his touring holiday; she also feels aggrieved that her husband and their 'careful chauffeur' should be 'treated like road-hogs' (p. 94). Margaret pointedly disagrees with Mrs Wilcox in this description of Henry: 'He was exceeding the speed limit, I conclude. He must expect to suffer with the lower animals' (p. 94). Speeding thus carried with it a sense of improper behaviour, as indicated by a remark in *The Times* for 1903 that road-hogs came 'from a class which possess money in excess of brains or culture'.[62]

If, for the Futurists, the speed of the car signified the push of modernity itself, others, such as Masterman, warned of the danger when 'life has become "speeded-up" to the motor-car level'.[63] Concern at what correspondence in *The Times* called 'motor tyranny' led the Government in 1905 to institute a royal commission on motoring, with Prime Minister Balfour telling the cabinet that 'public attention has of late been much drawn to the subject of motor cars. Numerous accidents have occurred with which these vehicles have been concerned, and the danger and annoyance caused by them have given rise to great complaint.'[64] Ironically, Balfour himself was one of those prominent motorcar-owners known to have broken the 20 mph speed limit.

The final worry attached to motorcars was the deterioration of the roads, and the extent of the dust produced by the increased volume of traffic. Dusty roads became a constant source of annoyance to those who disliked the car, and Forster's first description of a car journey is quick to utilise this contemporary phobia. Charles Wilcox drives Mrs Munt from the station at Hilton to the house at Howards End, dust blowing into her eyes in a car probably lacking a windscreen. Charles then stops at a village shop, and

> turned round in his seat, and contemplated the cloud of dust that they had raised in their passage through the village. It was settling again, but not all into the road from which he had taken it. Some of it had percolated through the open windows, some had whitened the roses and gooseberries of the wayside gardens, while a certain proportion had entered the lungs of the villagers. 'I wonder when they'll learn wisdom and tar the roads,' was his comment. (*HE*, p. 33)[65]

Charles's disregard for anything other than road quality looks forwards to the association of the Wilcox family with modernity and the idea of movement, whether literally, as here, or metaphorically, as in Margaret's view that 'Henry was always moving and causing others to move, until the ends of the earth met' (p. 323). Forster's attention to the dusty debris ignored by the Wilcoxes' rush to traverse space is very similar to one testimony on the distressing effects of motoring presented to the royal commission in 1906:

> All the plants under glass were spoiled, all the flowers were spoiled, all the strawberries and grapes were spoiled, and our health was injured. I had an inflamed throat all summer, and my eyes were very troublesome ... I had to get new typewriters ... in 1902 and I had to change them again this year, they got so gritty.[66]

Besides the harmful effect upon the body, it is noteworthy that this discourse around dust is part of an ongoing spatial conflict between the car as an emblem of the city and a countryside that is being 'spoiled' by the malignity of modern machinery. As Plowden notes, the 'conflict between motorists and others was partly an urban/rural one', with motorists being 'criticized for endangering relationships between the gentry and the peasantry'.[67] This city versus country opposition echoes throughout *Howards End*, and indicates how Forster's ambivalent view of the car is bound up with a variety of other representational spaces in his narrative.

The first description of Mrs Ruth Wilcox plays upon her embodiment of the pastoral values of place and which are opposed to the zip of the speeding motorcar that has just brought Mrs Munt to Howards End: 'She seemed to belong not to the young people and their motor, but to the house, and to the tree that overshadowed it. One knew that she worshipped the past' (p. 36).[68] Forster's descriptions of Mrs Wilcox continually identify her with a sense of Heideggerian dwelling threatened by the modernity of the motorcar. We are told later that 'clever talk' alarmed Mrs Wilcox, for 'it was the social counterpart of a motor-car, all jerks, and she was a wisp of hay, a flower' (p. 84). When we first see Ruth Wilcox she carries a wisp of hay, an emblem that looks forwards to the final cutting of the field at the end of the novel and re-emphasises her identification with the pastoral. Her husband, by contrast, suffers from hay-fever, as do all the male members of the Wilcox family (p. 268). Forster's use of the motorcar clearly employs a strongly gendered discourse, whereby women represent the supposed stability of rural places and the domestic home, with the Wilcox men representing the city, the car and modernity itself.[69] The male preserve of the smoking-room at the Wilcoxes' Ducie Street house, for example, resembles the motorcar because of their shared upholstery: the men 'smoked in chairs of maroon leather. It was as if a motor-car had spawned' (p. 167).[70]

At the novel's close Helen, commenting on the 'red rust' of suburbia encroaching on Howards End, notes that 'London is only part of something else, I'm afraid. Life's going to be melted down, all over the world' (p. 329). The city and the car exemplify this melting modernity, with the male Wilcoxes epitomising this dissolving existence. Early in the novel Helen is fascinated by the 'energy' (p. 37) of the Wilcoxes, and throughout the novel they are aligned with what Harvey terms the experience of 'time–space compression' during periods of capitalist expansion.[71] The Wilcoxes, deeply conservative in political terms, are clearly linked to the forces of technical and commercial modernisation that – like the energies of the car – are reshaping material spaces. Without people like the Wilcoxes, says Margaret, there 'would be no trains, no ships to carry us literary people about in, no fields even' (p. 177). But if the Schlegels have the cultural insight to reflect upon social space, it is the Wilcoxes who economically produce these spaces.

As Edward Said and Fredric Jameson have noted, Henry Wilcox's work for the Imperial and West African Rubber Company relates the novel to the production of imperial spaces abroad.[72] It is worth stressing, however, how this connection between the metropolis and imperial domains draws upon another kind of movement across space. In Henry's office Margaret views a map of Africa, 'on which the whole continent appeared, looking like a whale marked out for blubber' (p. 196). The rubber industry expanded greatly towards the end of the nineteenth century, with imports from British colonies forming the majority of such trade.[73] The explorers Stanley and Livingstone had both indicated to the rubber industry (which had previously relied on Amazon rubber) the importance of developing the trade in Central Africa.[74] One of the main uses for imports of rubber was in the production of pneumatic tyres to equip bicycles and motorcars. The rubber wheels of the Wilcoxes' motorcars were first invented in 1888 by John Dunlop, and they bring together the spaces of imperialism and the metropolis with its 'craze for motion'. The need for such materials was even noted in Parliament in 1910: 'The progress of electrical science, of motoring, and even of sports, has caused an enormous advance in rubber.'[75] Henry's work in the rubber industry shows how the text combines different spaces, and the textual transit in the novel between city and country echoes this wider geographical movement of imperialist trade, one that enables the Wilcoxes quite literally to travel by car from London to Howards End. My point here is not just to stress a particular historical experience of various spaces represented in the novel, but to argue that such a sense of modernist geographies profoundly enters into the narrative shape and style of Forster's text.

One example of this intrusion of social space into the novel is the journey which occurs in chapter 23, when the newly engaged Henry and Margaret are driven from his London office to visit Howards End. The incident is structured around spatial disorientation versus the fixity of place; it also contains ghostly, occluded images of the imperial spaces underpinning the journey. The drive itself, 'a form of felicity detested by Margaret', is not a success; Henry advises Margaret to look at the scenery from the car, but when she does so the landscape 'heaved and merged like porridge' (pp. 198, 199). The tourist gaze upon the countryside from the car, much praised by motor enthusiasts, is negated by the narrator's comment that 'Hertfordshire is scarcely intended for motorists', because its 'delicate structure particularly needs the attentive eye' (p. 198). The speed of the car thus thwarts this drive with a view, and Forster views the countryside missed, or which sinks into 'porridge', with a feeling of national space: 'Hertfordshire is England at its quietest ... it is England meditative' (p. 198). Here we see, yet again, the idea of connection across different spaces: from Henry's city office, with its imperial business, we move through London suburbs and across an essentially English pastoral space.

However, Forster's point is not just to show the interweaving of different spaces, but rather concerns the psychic consequences of spatial disturbance brought about by travel. The journey makes Margaret believe that she has 'lost all sense of space', only to regain it when she walks, for the first time, through the interior of Howards End: 'she recaptured the sense of space which the motor had tried to rob from her' (pp. 199, 201).[76] The chapter concludes with Margaret, prompted by the inner sanctum of the house, meditating upon this 'sense of space'. Interestingly, the interiorised narrative here replicates the sense of travelling between different spaces offered by the motorcar. Moving through the spaces of modernity can only be achieved, Forster suggests, from a position of fixity; but the crucial fact about Margaret's geographical imagining is the way that connections seem not to settle into a discourse of place. First, she contrasts the size of London with that of Howards End, commenting that 'ten square miles are not ten times as wonderful as one square mile' and that 'the phantom of bigness, which London encourages, was laid for ever when she paced from the hall at Howards End to its kitchen' (p. 201).[77] She then recalls what Helen had said about her proposed marriage to Henry: 'You will have to lose something' – a comment made while Helen was 'scrutinizing half Wessex from the ridge of the Purbeck downs' (p. 201) in the chapter whose action takes place in Swanage. Margaret, surveying the internal space of Howards End rather than the rolling English hills, now disagrees, for 'she would double her kingdom by opening the door that concealed the stairs' (p. 202). Personal emotions are conveyed through spatial metaphors that draw upon the significance of actual material spaces: the 'kingdom' of England is transferred into the domestic 'kingdom' that Margaret might possess at Howards End. Margaret's fear of losing her sense of space is also transformed here: she will not lose anything but, in the manner of the imperialist, will gain more space. However, this is not material space but a psychic space represented in her 'kingdom' of the house. The inner life of the spirit and the external life of the material world are one of the key binaries the novel addresses; my point here – one which has not been sufficiently noticed before – is that the opposition owes much to the geographical imagination of the text, and that material spaces, those external to the text, form a key part in the transformation of the metaphorical inner space of the novel itself. As Homi Bhabha comments, *à propos* Henry James: 'The recesses of the domestic space become the sites for history's most intricate invasions. In that displacement, the borders between home and world become confused; and … the private and public become part of each other'.[78] Reading the novel with this kind of emphasis reveals quite clearly the specific form of Forster's version of modernism, where 'ambiguous continuities', in Lefebvre's terms, exist between the multiple spaces of modernity.

It is significant that Forster extends this series of transforming spatial metaphors, linking this gendered space of kingdom back to the imperial spaces suggested by Henry's work in London: 'Now she [Margaret] thought of the map of Africa; of empires; of her father; of the two supreme nations, streams of whose life warmed her blood, but, mingling, had cooled her brain. She paced back into the hall, and as she did so the house reverberated' (p. 202). Here Forster, collapsing his narrator into that of Margaret, strains to indicate the relations between different spaces, but the dominant sense of this passage – itself as close as Forster comes to a stream of consciousness narration – is ultimately a confused one: for what are the connections between a house, English hills, the two nations of England and Germany, a map of Africa, and the gendering of these spaces? Margaret's 'sense of space' reveals only a cooled brain, a rather odd image indicating an inability to fit together these *topoi* in a significant fashion. The narrator is clearly aware that important relations exist between these diverse spaces, but is unable to quite decide what they are, or how they are associated. One option might have been for the novelist to suggest that the real connections between such locations can exist only in motion, in provisional though pertinent sets of relations between country, city, Empire and house; but that strategy would yield too much to the nomadic claims of the motorcar. Forster craves a sense of connection rooted in fixed symbols and images – in de Certeau's terms, a map; but the thrust of his narrative and the modernity it encounters stresses the *process* of connection, of touring through the modern, rather than a representation of space in a finished and reified form.

Another way to put this point would be to say that here Forster's novel keeps brushing up against a modernity it can, at times, represent, but which it cannot truly – because it does not desire to – understand or embrace. Margaret's empty encounter with space here recalls a comment of Buzard in his analysis of Forster's Italian travel novels, *A Room with a View* and *Where Angels Fear to Tread*. Discussing how English characters in these novels are always looking for an 'authentic' experience of Italian otherness, only to somehow miss that which is truly 'Italian', Buzard comments: 'Forster's characters repeatedly enact a failed encounter with the "real" which they believe themselves to have met.'[79] Something of this perception is also evident in *Howards End*. Margaret's failure here lies in her encounter with the house at Howards End; she believes that this house is the 'real' place, in de Certeau's terms, from which other spaces can be mapped. This view is, ultimately, a retreat from connecting diverse spaces together because it is a place of fixity and not flux, an emblem of an enclave from modernity and not a vantage point from which to survey it.

This problematical encounter with space is repeated, in the next chapter, when Margaret stays at Howards End. Henry has to travel to London by car early the next day, and this reminder of the metropolis takes Margaret on a disorientating mental voyage once again: 'Once more she lost the sense of space; once more trees, houses, people, animals, hills, merged and heaved into one dirtiness, and she was at Wickham Place' (p. 204). Margaret's psychic space is overwhelmed by that of the motorcar, and the elliptical narrative here – it takes some attention for the reader to realise 'she was at Wickham Place' only in a metaphorical sense – is yet another instance of Forster's shift towards a modernist style where metaphorical and material spaces merge and overlap.

The next paragraph has Margaret repeat her engagement with space:

> Her evening was pleasant. The sense of flux which had haunted her all the year disappeared for a time. She forgot the luggage and the motor-cars, and the hurrying men who know so much and connect so little. She recaptured the sense of space, which is the basis of earthly beauty, and, starting from Howards End, she attempted to realize England. She failed – visions do not come when we try, though they may come through trying. (*HE*, p. 204)

The moment appears, initially, as another attempt, as at Swanage, to understand national spaces, to convert the abstractions of space into the known realities of place by the process of 'connection'. The 'flux' of the car is replaced by the fixities of place, once again the house. Interestingly, the exercise fails, although Margaret is filled with 'an unexpected love of the island ... connecting on this side with the joys of the flesh, on that with the inconceivable'. This strange set of emotions 'had certainly come through the house and old Miss Avery. Through them: the notion of "through" persisted: her mind trembled towards a conclusion which only the unwise have put into words' (p. 205). Miss Avery is the housekeeper who, on the previous evening, had disturbed Margaret in the house and, tellingly, had mistaken her for Ruth Wilcox. Characteristically, Forster refuses to spell out the precise significance of this incident; but it clearly involves trying to cognise the spaces of modernity and, furthermore, attempting to grasp the idea of moving 'through' them by new forms of transport such as the motorcar. It is also a fine textual example of the pressure of David Harvey's 'time-space compression': 'through' here also signifying a movement through the years, through the past owners of the house itself.

Flux 'haunts' Margaret, and it is this agitated quality which she is unable to 'realize'. Links, for Margaret, cannot be made between loci that refuse, even temporarily, to stay still. More than this, we might say that Margaret's failure here stems from the very nature of *connect* as a word:

for it is a verb, a term of action, more allied to the process of flux associated with motors and modernity. In a sense, it is the very experience of modernity that Forster is representing here; an experience not of specific *places* (Howards End, London, England) but of the processes of spatial production, marked here by the mere concept of 'flux' and the notion of 'through' to which the novelist can gesture, but which he cannot literally represent. As Jameson comments on Forster, it is in these sorts of moments that we can discern the 'modernist style' of the novel, where 'an appearance of meaning is pressed into the service of the notation of a physical perception'.[80] Significantly, in the original manuscript for the novel the word 'realize' was originally 'visualize', emphasising a form of tourist gaze rather than a cognitive connection.[81] Perception of the house and then of England can only be rendered meaningful by the puzzling notions of *through*, *realize* or *connection*.

The motorcar, then, is an ambivalent symbol in *Howards End*, for although Forster is consistent in his censure of it for encouraging the 'craze for motion' of a nomadic modern life, it is this very quality of moving *through*, as in the through-traffic brought by the motorcar, that becomes important when Forster's characters try to 'realize' and understand the contemporary world. The car represents a sense of flux disrupting an older idea of the *genius loci*, and which pushes the novel, perhaps against its author's wishes, towards a more developed modernist narrative. Looking back, later in his life, Forster commented mournfully upon the motif of movement in his work: 'Howards End is a hunt for a home. India is a Passage for Indians as well as English. No resting-place.'[82] In a text like *Howards End* we see an interplay between flux and form, the hunt and the home, with Forster unsuccessfully wishing to use literary and cultural form as a bulwark against the disorientations of modernity. Forster's strategy fails because modernity as flux always seems to burst through the containing strategies of literary form. The motorcar, then, becomes the symbol of narrative connection that ultimately subverts any structure of balanced binaries in the novel.

By connecting Forster's literary discourse to some historical discourses surrounding motoring I have shown how the text is best understood in the context of a spatial history of the experience of modernity. I have drawn attention to the diverse material spaces that Forster's novel represents – rooms, houses, cities, suburbs and nations – and I have linked these social spaces with psychic space, most notably shown in Margaret's problematical encounter with 'flux'. The motorcar seems also to represent quite directly a 'struggle over geography' that involved conflict over the new spaces of modernity, such as the relation between city, suburb and countryside.[83] Another key set of spaces the novel explores is that of the links between the metropolitan motorcar and the economic flows of imperial trade. Indeed, Forster makes an explicit correlation between the motorcar and the figure of 'the Imperial ... ever in motion ... he prepares the way for cosmopolitanism' (*HE*, pp. 314–15). Reading the novel in this spatial, or literary geographical, fashion produces a considerably more interesting novel than an interpretation that stresses only Forster's failure to accommodate the more familiar modernist techniques in the novel, such as corrosive irony or a stream of consciousness narration.

I finish with one final image of the battle between flux and form, space and place, in Forster's novel. Musing upon the opposition between a life devoted to the world of the spirit and one concerned with material gain, Margaret recalls, and dismisses, Aunt Juley's claim that the truth of life lies halfway between these two ideas: 'No; truth, being alive, was not halfway between anything. It was only to be found by continuous excursion into either realm, and, though proportion is the final secret, to espouse it at the outset is to ensure sterility' (*HE*, pp. 195–6). This gives the lie to those readings of the novel that focus solely upon the neat structuring features of the narrative. The spatial imagination of the excursions of the motorcar is the key to an interpretation that finds in the novel a different, and ongoing, attitude towards modernity. The modernism of *Howards End*, I would suggest, exists within its 'continuous excursion' rather than its sense of 'proportion', which remains, in that text at least, a secret whose hidden spaces it is our function as readers to discover and connect.

Chapter 3 picks up this equation of modernity with motion and shifts the scene to a different set of spaces and a different form of metropolitan transport: the underground tube train.

Notes

1 Few modernist texts locate their composition in this way. Perhaps the most famous exception is the end of Joyce's *Ulysses*: 'Trieste-Paris-Zurich, 1914–21'. Whereas Forster's single name indicates his allegiance to a unity of place, Joyce's European triumvirate indicates a migratory cosmopolitanism that embraces space more than place.

2 See, for example, Peter Keating, *The Haunted Study: A Social History of the English Novel 1875–1914* (London, Fontana, 1991), pp. 325–7; and Lyn Pykett, *Engendering Fictions: The English Novel in the Early Twentieth Century* (London, Edward Arnold, 1995), pp. 117–22.

3 E. M. Forster, *The Manuscripts of* Howards End, ed. Oliver Stallybrass (London, Edward Arnold, 1973), Abinger edition, vol. 4a, p. 275.

4 In a number of his early writings Forster was drawn to the concept of the *genius loci*, the sense of a body of personal associations or inspirations intrinsically linked to a place. He noted: 'One of my novels, *The Longest Journey*, does indeed depend from an encounter with the genius loci, but indirectly, complicatedly.' See E. M. Forster, 'Introduction', *Collected Short Stories* (Harmondsworth, Penguin, 1954), p. 6.
5 Daniel Schwarz, *The Transformation of the English Novel 1890–1930* (Basingstoke, Macmillan, 1989), p. 118.
6 Franco Moretti discusses the 'phenomenology of the border' in *Atlas of the European Novel 1800–1900* (London, Verso, 1998), pp. 33–47.
7 Howards End was based directly upon Rooksnest, a house near Stevenage, Hertfordshire, where Forster lived from the age of 4; see Nicola Bauman, *Morgan: A Biography of E. M. Forster* (London, Hodder & Stoughton, 1993), pp. 30–43.
8 Elizabeth Langland, 'Gesturing towards an open space: gender, form and language in *Howards End*', in Jeremy Tambling, ed., *E. M. Forster: New Casebook* (Basingstoke, Macmillan, 1995), p. 81.
9 Perry Meisel, *The Myth of the Modern: A Study in British Literature and Criticism after 1850* (New Haven and London, Yale University Press, 1987), p. 169; Fredric Jameson, *Modernism and Imperialism* (Derry, Field Day Pamphlet, 1988) and Brian May, *The Modernist as Pragmatist: E. M. Forster and the Fate of Liberalism* (Columbia and London, University of Missouri Press, 1997). For a re-evaluation of Forster in the light of queer theory see Joseph Bristow, *Effeminate England: Homoerotic Writing after 1885* (Buckingham, Open University Press, 1995), ch. 2.
10 I am using the term 'attitude towards modernity' in the way suggested by Michel Foucault in his essay 'What is Enlightenment?' in *The Foucault Reader*, ed. Paul Rabinow (London, Penguin, 1984), p. 39.
11 Richard Sennett, *Flesh and Stone: The Body and the City in Western Civilization* (London, Faber, 1994), p. 354.
12 Ruskin had made this point in relation to the railways in the nineteenth century: 'The railroad is in all its relations a matter of earnest business, to be got through as soon as possible. It transmutes a man from a traveller into a living parcel': John Ruskin, 'The lamp of beauty' in *The Seven Lamps of Architecture* (London, George Allen, 1907), p. 220.
13 The phrase is Marx's, famously developed in relation to modernism by Marshall Berman in his book *All that Is Solid Melts into Air: The Experience of Modernity* (London and New York, Verso, 1983).
14 Lewis Mumford, *The City in History: Its Origins, its Transformations and its Prospects* (Harmondsworth, Penguin, 1966), p. 13.
15 This rebuilding does indeed happen later in the book (*HE*, p. 253).
16 In the five years previous to *Howards End*, Forster had also published *Where Angels Fear to Tread* (1905), *The Longest Journey* (1907), and *A Room with a View*

(1908). After *Howards End* he completed *Maurice* in 1914 but did not publish it, and thereafter published only *A Passage to India*, some ten years later.

17 Earlier the link between spatiality and money is introduced by Margaret's comment to Helen that 'You and I and the Wilcoxes stand upon money as upon islands' (p. 72).

18 John Carey suggests that the term 'suburban' was 'distinctive in combining topographical with intellectual disdain'. John Carey, *The Intellectuals and the Masses: Pride and Prejudice among the Literary Intelligentsia, 1880–1939* (London, Faber, 1992), p. 53. For a general overview of how popular fiction represented the suburbs in this period see Kate Flint, 'Fictional suburbia', in Peter Humm, Paul Stigant and Peter Widdowson, eds, *Popular Fictions: Essays in Literature and History* (London, Methuen, 1986).

19 See, for example, Pound's poem *Commission*, from *Lustra* (1916), which calls upon his verse to 'Speak against unconscious oppression' and 'Go to the bourgeoise who is dying of her ennuis,/ Go to the women in suburbs'. Ezra Pound, *Collected Shorter Poems* (London, Faber, 1984), p. 88.

20 Forster read with enthusiasm the articles which formed Masterman's book (see Bauman, *Morgan*, p. 219). Crosland's vituperative book contains a chapter entitled 'The female suburban' in which he refers to the 'dense and torpid mind of the female suburban'. T. W. H. Crosland, *The Suburbans* (London, John Long, 1905), p. 48, and claims that '[w]omen's rights were first broached, and bloomers first worn, in Suburbia' (p. 69). Ironically, Forster's Weybridge home was described by his biographer P. N. Furbank as a 'commonplace three-storeyed suburban villa'. See P. N. Furbank, *E. M. Forster: A Life*, vol. 1: *The Growth of the Novelist 1879–1914* (London, Secker & Warburg, 1977), p. 119.

21 C. F. G. Masterman, *The Condition of England* (London, Methuen, 1909), p. 70.

22 Crosland, *The Suburbans*, pp. 79, 88.

23 Flint, 'Fictional suburbia', p. 113.

24 Quoted in Flint, 'Fictional suburbia', pp. 113–14.

25 This was a point acknowledged by Forster himself in an interview in 1958 (see *HE*, p. 14).

26 The classic contemporary discussion of the image of the 'abyss' is that by Jack London, *People of the Abyss* (1903); for critical discussions of the trope see John Marriott, 'Sensation of the abyss: the urban poor and modernity', in Mica Nava and Alan O' Shea, eds, *Modern Times: Reflections on a Century of English Modernity* (London, Routledge, 1996); and Joseph McLaughlin, *Writing the Urban Jungle: Reading Empire in London from Doyle to Eliot* (Charlottesville and London, University of Virginia Press, 2000).

27 Forster does, however, acknowledge his inability to represent certain sections of society: 'We are not concerned with the very poor', he writes at the start of chapter 6. 'They are unthinkable ... This story deals with gentlefolk' (*HE*, p. 58).

28 David Trotter, *The English Novel in History 1895–1920* (London, Routledge, 1993), p. 83.

29 John Burnett, *A Social History of Housing 1815–1985*, 2nd edn (London, Methuen, 1986), p. 155.

30 I have ignored Leonard's walk into Surrey, which is not represented by the narrator, and the various references to visits to Germany, Italy and Nigeria, which are not represented at all. The brief mentions of Tibby's life as a student at Oxford are perhaps more significant, if only for Forster's comment on 'how faint [was] its claim to represent England' (*HE*, p. 251).

31 Compare Moretti's comments on the narrow Englishness of Jane Austen's geographical world: Moretti, *Atlas of the European Novel*, pp. 13–18.

32 Rather clumsily, the chapter does contain a foreigner, a German visitor staying with Mrs Munt, who is shown the view over Poole harbour.

33 H. J. Mackinder, *Britain and the British Seas*, 2nd edn (Oxford, Clarendon Press, 1907), p. 258.

34 The use of pronouns indicates that Forster's narrator is English, with an English audience in mind.

35 For discussion of the significance of the home rule debate in this period see, among others, D. G. Boyce, 'The marginal Britons: the Irish', in Robert Colls and Philip Dodd, eds, *Englishness: Politics and Culture 1880–1920* (London, Croom Helm, 1986) and *Nationalism in Ireland*, 2nd edn (London, Routledge, 1991), ch. 9; and R. F. Foster, *Modern Ireland 1600–1972* (London, Penguin, 1989), pp. 397–428.

36 Franco Moretti, *Atlas of the European Novel*, p.46.

37 John Lucas, 'Discovering England: the view from the train', *Literature and History* 6: 2 (autumn 1997), 38.

38 Richard Overy, 'Heralds of modernity: cars and planes from invention to necessity', in M. Teich and Roy Porter, eds, Fin de Siècle *and its Legacy* (Cambridge, Cambridge University Press, 1990), p. 54.

39 John Stevenson, *British Social History 1914–45* (Harmondsworth, Penguin, 1984), p. 27.

40 Stephen Kern traces this ambivalence in relation to the speed of the car in *The Culture of Time and Space, 1880–1918* (Cambridge, MA, Harvard University Press, 1983), ch. 5.

41 Algernon E. Berriman, *Motoring: An Introduction to the Car and the Art of Driving it* (London, Methuen, 1914), p. 32.

42 C. W. Brown, *ABC of Motoring* (London, Henry J. Drane, 1909), pp. 7, 8.

43 R. P. Hearne, *Motoring* (London, Routledge, 1908), p. 6.

44 See Peter Nicholls, *Modernisms: A Literary Guide* (Basingstoke, Macmillan, 1995), pp. 172–3; and Giovanni Cianci, 'Futurism and the English avant-garde: the early Pound between Imagism and Vorticism', *Arbeiten aus Anglistik und Amerikanistik* 1 (1981), 3–39.

45 F. Marinetti, 'The founding and manifesto of Futurism', in Umbro Apollonio, ed., *Futurist Manifestos* (London, Thames & Hudson, 1973), p. 21.

46 Hearne, *Motoring*, pp. 6 and 5.

47 Edith Wharton, *A Motor Flight Through France* (1908) (London, Picador, 1995), p. 17.
48 Berriman, *Motoring: An Introduction*, p. 30. This point is also made by Wharton.
49 William Plowden, *The Motor Car and Politics in Britain* (Harmondsworth, Penguin, 1973), p. 15.
50 Plowden, *Motor Car and Politics*, p. 65.
51 Quoted in Peter Roberts, *The Motoring Edwardians* (London, Ian Allan, 1978), p. 46.
52 For discussion of this issue see the essays in Colls and Dodd, *Englishness: Politics and Culture*, as well as Jose Harris, *Private Lives, Public Spirit: Britain 1870–1914* (Harmondsworth, Penguin, 1994).
53 Another close friend of Forster, Goldsworthy Lowes Dickenson, published two articles which attacked the growth of motoring: 'Motoring', *The Independent Review* (January 1904), 578–89 and 'The motor tyranny', *The Independent Review* (October 1906), 15–22.
54 C. F. G. Masterman, *The Condition of England* (London, Methuen, 1909), p. 65.
55 Plowden, *Motor Car and Politics*, p. 23.
56 This point is made in Overy, 'Heralds of modernity', p. 73.
57 See Plowden, *Motor Car and Politics*, p. 5; for figures on accidents see Kern, *Culture of Time and Space*, p. 113.
58 It might also be said that, as Wright notes, Leonard's death is also partly caused by the motorcar, as he is overtaken by Charles in his car when walking from Hilton to Howards End. See Anne Wright, *Literature of Crisis 1910–22* (London, Macmillan, 1984), p. 31.
59 Plowden, *Motor Car and Politics*, p. 4.
60 Plowden, *Motor Car and Politics*, pp. 42–3.
61 See Hearne, *Motoring*, p. 5. In 1910 the land speed record was set by a German Benz at 131.72 mph; see Anthony Bird, *Early Motor Cars* (London, George Unwin, 1967), p. 166.
62 Quoted in Plowden, *Motor Car and Politics*, p. 27.
63 Masterman, *Condition of England*, p. 23.
64 Quoted in Plowden, *Motor Car and Politics*, p. 58.
65 Ironically, it was only in 1910 that roads were first tarred; see Plowden, *Motor Car and Politics*, p. 85.
66 Quoted in Plowden, *Motor Car and Politics*, p. 60.
67 Plowden, *Motor Car and Politics*, pp. 23–4.
68 Kern notes how the phenomenon of speed associated with the car produced a reactive nostalgia for the perceived slowness of the past: see *Culture of Time and Space*, pp. 129–30.
69 Trilling, discussing how the novel represents the conflict between the sexes, notes that the car is 'the totem of the Wilcox males': Lionel Trilling, *E. M. Forster: A Study* (London, Hogarth, 1969), p. 109. Gender categories also help

organise other spaces in the novel. Early on we learn that the Schlegel's London home is 'a female house', and is contrasted with the masculine abode of the Wilcoxes (*HE*, p. 56). When Margaret and her pregnant sister Helen are reunited in Howards End – a house identified with the first Mrs Wilcox – they 'defend' this feminine space from Henry Wilcox and the male doctor: 'A new feeling came over [Margaret]: she was fighting for women against men. She did not care about rights, but if men came into Howards End it should be over her body' (p. 283). The Schlegels' furniture fits perfectly the interiors of the house, although men have 'spoilt' one room 'through trying to make it nice for women' (p. 290).

70 A point suggested by Wright, *Literature of Crisis 1910–22*, p. 31.

71 David Harvey, *The Condition of Postmodernity* (Oxford, Blackwell, 1990), pp. 240–2.

72 See Edward Said, *Culture and Imperialism* (London, Vintage, 1994), p. 77; and Jameson, *Modernism and Imperialism*. I'm using 'production' here in the sense that Lefebvre does: all space is the result of human activity, and bears the marks of this productive activity (*PS*, pp. 84–5).

73 William Woodruff, *The Rise of the British Rubber Industry During the Nineteenth Century* (Liverpool, Liverpool University Press, 1958), p. 38.

74 Woodruff, *British Rubber Industry*, p. 39.

75 Bernard Porter, *The Lion's Share: A Short History of British Imperialism 1850–1983*, 2nd edn, (London, Longman, 1984), p. 222.

76 This fear of a loss of space also occurs in Forster's early short story 'The machine stops' in *Collected Short Stories*, p. 125.

77 Also see *HE*, p. 43 for discussion of the 'bigness' of imperialism: city and Empire are linked here because both expand over space.

78 Homi K. Bhabha, *The Location of Culture* (London and New York, Routledge, 1994), p. 9. Bhabha is discussing *The Portrait of a Lady*.

79 James Buzard, *The Beaten Track: European Tourism, Literature and the Ways to 'Culture' 1800–1918* (Oxford, Clarendon Press, 1993), p. 314. It is interesting to note that although Charles Wilcox honeymoons in Italy, what 'he enjoys most is a motor tour in England' (*HE*, p. 81).

80 Jameson, *Modernism and Imperialism*, p. 15.

81 Forster, *The Manuscripts of* Howards End, p. 208.

82 Forster, 'A view without a room' (1958), appendix to *A Room with a View* (Harmondsworth, Penguin, 1978), p. 232.

83 Said, *Culture and Imperialism*, p. 6.

3 Imagist travels

In a 1909 article entitled 'Modern poetry' for the little magazine *The Thrush*, Ford Madox Ford defined the conditions required for a renaissance in modern poetry. It would appear only when poets who 'shut themselves up in quiet book-cabinets' or who 'dream for ever of islands off the west coast of Ireland' emerge and 'get it into their heads to come out of their book-closets and take, as it were, a walk down Fleet St., or a ride on the top of a bus from Shepherd's Bush to Poplar. I am using, of course, these peregrinations metaphorically.'[1] Ford continues his use of transport as a trope for modernity, when he compares the view from his window then, in the early 1900s, with what he would have seen a century earlier, looking out on a 'great western-going highway'. He prefers the contemporary view, where one sees

> innumerable motes of life in a settled stream, in a never-ceasing stream, in a stream that seems as if it must last for ever ... And all these impressions are so fragile, so temporary, so evanescent, that the whole stream of life appears to be a procession of very little things, as if, indeed, all our modern life were a dance of midges.

He explains this conception of modern life in a way that clearly indicates the experience of modern transport: 'We know no one very well, but we come into contact with an infinite number of people; we stay nowhere very long, but we see many, many places.'[2] Modern life is thus defined in terms of an incessant movement through urban space, a psychic movement that is simultaneously social and cultural, metaphorical and literal.

This chapter suggests that we take Ford's metaphorical bus journey rather more literally as a route into thinking about the development of modernist poetry, or, in this case, the poetry of Imagism, one of the most significant of early Anglo-American modernist formations.[3] For around the time that Forster was tackling the flux of the motorcar in *Howards End* Imagism was being formulated as an aggressively contemporary movement in verse, one designed to register the new sensations of modernity. Imagism listened attentively to Ford's advice and took the bus and underground tube train as the setting for a number of its poems. A cluster of Imagist poems, found in five anthologies between 1914 and 1930, demonstrates how the quotidian experience of moving around the metropolis provided an important impetus to some of the experimental forms of modernist writing. By examining the aesthetic practices of this key formation within early Anglo-American modernism we can determine a set of experiences of space and geography different from those found in Forster. One significant distinction is the kind of intersubjective visual relations that dominate city life and which inform the theory of 'the image'. In this chapter the focus upon social space is extended to consider theories of the gaze, Lefebvre's 'logic of visualisation', and the gendered nature of the spatial relations of looking.

One example of how the geography of the city is connected to questions of gender and looking can be found in T. E. Hulme's 'The Embankment (The fantasia of a fallen gentleman on a cold, bitter night)', a poem not concerned with transport:

> Once, in a finesse of fiddles found I ecstasy,
> In a flash of gold heels on the hard pavement.
> Now see I
> That warmth's the very stuff of poesy.
> Oh, God, make small
> The old star-eaten blanket of sky,
> That I may fold it round me and in comfort lie.[4]

This early Imagist poem, first published in 1909, concerns an experience of urban 'ecstasy' prompted by the visual 'flash' of a pair of 'gold heels'. The final image, likening the starry sky to an old blanket, develops from the insight that 'warmth's the very stuff of poesy'. But this point, and the image of warmth embodied in wrapping the sky around oneself, is prompted only by the initial visual experience of the discovery of 'ecstasy'. From the poem's subtitle we can deduce that 'gold heels' refers to a woman perceived while walking along the Embankment in London. Relations of a sexualised nature are suggested, maybe not consummated since this is a 'fantasia', but certainly crystallised around the 'flash' of a gentleman's gaze. 'Fallen' carries the hint of sexual sinning, perhaps indicating that the woman is a prostitute; the fall also matches the downward movement of the man's gaze onto the fetishised female heels.[5]

The poem, then, offers an interesting illumination of the urban experience as represented in the modernist period: the spatial relations of looking are interwoven with sexual and gender relations. Transcending the urban context is achieved by a moment of libidinised visual pleasure, transformed into a literary text. Before examining how space was negotiated in the poems of Imagism, it is useful to remember how the city and its spaces have re-emerged in contemporary discussion of the nature of postmodernity.

In a series of influential texts on postmodernism Fredric Jameson has outlined the need for a new 'cognitive mapping' of postmodernist urban space. Jameson argues that we are 'in the presence of something like a mutation in built space' which we are unable to comprehend completely. This is because 'we do not yet possess the perceptual equipment to match this new hyperspace'. Our sensual understanding of the built environment of the postmodern city is still lodged in the world of modernism: 'our perceptual habits were formed in that older kind of space I have called the space of high modernism'. Production of a 'cognitive map' able to guide and help comprehend this new 'hyperspace' requires the growth of our perceptual capabilities through an expansion of 'our sensorium and our body'.[6] Equipped with refined bodily organs we can then attempt to develop our self-consciousness as postmodern subjects in this new spatial environment. Jameson's argument here can be condensed into the idea that at present there exists a gap between the social relations of postmodernity and the spatial relations of our senses. Our perceptual sense of space is a modernist one; our social situation is a postmodernist one. In order to comprehend the social level we require a re-orientation at the level of our understanding of space.

A close reading of Imagist poems about transport suggests a rather different picture from that offered by Jameson. In particular, Jameson's model is a little reductive in its reading of modernist space, and is also insufficiently attentive to the gendering of space. Reading Imagism through postmodernist eyes draws attention to how far we rely upon sorts of socio-spatial relations similar to those represented in modernism. Issues of gender and visual space raised by the Hulme poem indicate that Jameson's understanding of our 'perceptual habits' is inadequate as it stands. The social spaces of the city – whether today or in modernism – are sites of conflict over gender and sexuality. Feminist cultural theory has recognised this point, especially in relation to urban space. Discussion has focused upon the gendering of the intimacies proffered by city spaces. In what might be called the great *flâneur* debate, critics such as Susan Buck-Morss, Griselda Pollock, Elizabeth Wilson and Janet Wolff have considered the availability or unavailability of city space for women from the nineteenth century onwards.[7] Griselda Pollock, for example, links the 'intimate space' of looking between men and women with the social spaces of the city in which such gazes occur, and attempts to illustrate how the formal experiments of modernist art are informed by these spatial relations. For this argument she draws upon two sources: psychoanalytic models of the 'male gaze', where women are fixed in ocular images as part of a masculine defence against castration anxiety; and the idea of the male *flâneur*, that city-dweller first noted by Baudelaire and subsequently described by Walter Benjamin as the archetype of modernity. The *flâneur*, writes Pollock, 'embodies the gaze of modernity which is both covetous and erotic'.[8] He is free to roam public city space with a 'detached observing gaze, whose possession and power is never questioned as its basis in the hierarchy of the sexes is never acknowledged'.[9] Quoting Janet Wolff in support of her view, Pollock argues that there can be no *flâneuse*, for women 'were never positioned as the normal occupants of the public realm. They did not have the right to look, to stare, scrutinize or watch.'[10] Women's spaces were predominantly those of private domestic interiors, like that of the first Mrs Wilcox in *Howards End*, while unaccompanied women in public places were taken to be prostitutes. It was these women, working the spaces of the modern city, who were figured as the objects gazed at by the *flâneur* during his metropolitan excursions.[11] T. E. Hulme's ecstatic glance at a woman's 'gold heels' is a good example of the habits of the *flâneur*.

This analysis by Pollock and Wolff has, however, been perceptively critiqued by Elizabeth Wilson. Rather than rehearse Wilson's arguments I want to extend the debate from the *flâneur* to the *voyageur*, as a way of specifying historically the socio-spatial relations obtaining in metropolitan London and, to a lesser extent, Paris in the period of the early Imagist poems. If transport is, as Forster noted, a 'forcing-house for the idea of sex', then it is important to discuss the ways in which this experience of space differs from that of the ambulant *flâneur*.[12] Transport is also, as I have already argued, an integral part of the 'time–space compression' of capitalist modernisation noticed throughout this period.

Before discussing the modernist spaces of transport, one point in Wilson's brilliant account of the *flâneur* bears emphasis because it suggests a convergence between these two interpretations of space. Wilson argues that the *flâneur* is actually a fiction, being nothing 'but an embodiment of the special blend of excitement, tedium and horror aroused by many in the new metropolis, and the disintegrative effect of this on the masculine identity ... He is a ... shifting projection of angst rather than a solid embodiment of male bourgeois power'. The 'male gaze' of the *flâneur* is an anxious response to the presence, rather than the absence, of women in the modern city, ultimately representing 'masculinity as unstable, caught up in the violent dislocations that characterized urbanization'.[13] It is the sexual and social forces of urban spatial relations that produce the nature of masculine looking in modernism, not simply psychic structures of dominance. The 'giddy space' of the city, argues Wilson, is 'too open' to be subject to any simple masculine manipulation; unbridled urban sensations cause the symptomatic pressure of disorientation, Jameson's lack of a 'cognitive map' or Forster's 'civilization of luggage'. The only defence against this destabilisation – the modernity where 'all that is solid melts into air' as a consequence of time–space compression – is to try to solidify and fix either oneself or other subjects. 'One such attempt', suggests Wilson, 'may be the representation of women in art as petrified, fixed sexual objects.'[14] It is, then, no surprise to find, as in Forster's 'forcing-house', that the 'giddy space' of urban transport staged in Imagist poetry replays this sexual and textual dialectic of fixity and dissolution. Male Imagism is marked by attempts to root one's self in the modernist maelstrom, often by fixing others in the spaces of one's poems. One site of these tussles is that of the London Underground. The tube condenses the two forms of space discussed so far: the spaces of newly modernised urban landscapes and the narrower spaces of the intimate relations between people when travelling. To these can be added a third conception of space: the representational spaces, in Lefebvre's terms, of poetic texts in which the forces of those other spaces are registered.

The London Underground

'Transportation is civilisation', trumpeted Ezra Pound in 1917, quoting from Kipling, and marking his disagreement with Forster's gloomy 'civilization of luggage'. Anything that interferes with traffic is an evil, claimed Pound, and a tunnel, such as one under the Channel that would connect London and Paris, is 'worth more than a dynasty'.[15] Imagism welcomed transport as a modern sensation *par excellence*, believing 'passionately in the artistic value of modern life', as the 1915 Imagist anthology proclaimed.[16] Taking your pet turtle for a walk in the city, as the *flâneur* did in 1840s' Paris to show distaste for the increased pace of life, was an option no longer available in a city like London in 1914, the year of the first Imagist anthology. As Susan Buck-Morss states: 'For the *flâneur*, it was traffic that did him in.'[17] But, for the Imagist, urban traffic offered many new perceptual possibilities. Imagism sought to represent in textual space the new urban spaces of capitalist modernisation.[18]

Discriminations between different cultural experiences offered by modern transport are also noticeable in non-Imagist poetry of the pre-war period. 'That railways are inadequate appears/ Indubitable now.'[19] These are not the words of a frustrated commuter in 2002 but those of 1890s' poet John Davidson in 'The Testament of Sir Simon Simplex Concerning Automobilism', a poem first published in *The New Age* in 1908. 'Simplex' was a brand name briefly used by the Mercedes motor company in the early years of the twentieth century. Davidson's poem contrasts two modes of transport and finds the political significance of cars preferable to that of trains. Railways are condemned for being 'democratic, vulgar, laic' because they marshal together all classes and sections of society: 'Bankers and brokers, merchants, mendicants,/ Booked in the same train like a swarm of ants'. Motorcars, however, emphasise the individual over the mass, for although 'the train commands, the automobile serves'. The 'privacy and pride' of the car expresses the 'Will to be the Individual' rather than the 'Will to be the Mob' inherent in rail travel. Davidson's debt to Nietzsche is very apparent in this poem, and it is interesting to note how modes of transport represent not only political, or semi-philosophical, points of view, but are also associated with a sense of modernity. If railways are negatively identified with democracy and socialism ('The socialistic and the railway age/ Were certainly coeval'), then both the mode of travel and the political ideal are seen as outdated:

> I call Democracy archaic, must
> As manhood suffrage is atavic lust
> [...]
> whose analogue
> In travel was the train, a passing vogue.

The car, however, looks forward to a new age of individuality, freed from antiquated notions of equality. With the car, 'A form, a style, a privacy in life/ Will reappear', and this new quality of experience will, as with much in English culture, be linked to a sense of the past: 'Now with the splendid periods of the past/ Our youthful century is proudly linked.' Davidson, I take it, would clearly not have taken Ford's bus to Poplar.

Although formally this poem belongs to the urban ballads popular with Davidson and others in the 1890s, it is modernist in the way that technology is associated with a distinctive experience of modern life itself. Davidson's paean to the pleasures of the automobile also points to the contemporary political meanings of the new machinery, with the car representing a rampant individuality found repugnant by Forster in *Howards End*. However, other forms of modern transport brought the modernist artist, as Davidson noted of the railways, into close proximity with many other constituencies. Travelling by tube certainly dragged poets out of their sealed-up book closets.

One of the key examples of modernisation in Edwardian London was the recently completed network of underground railways. These were praised by the Futurist poet Marinetti in 1912 for providing 'a totally new idea of motion, of speed', one which he preferred to the Paris Metro.[20] This interest in the technical modernisation of the railways may have influenced the Imagists, but the latter group never quite eulogised machinery in the manner of the Italian coterie. For Imagism transport represented a modern world redolent with anxieties as well as mechanical delights. Hulme, in 1914, may have argued that modern art was moving towards 'lines which are clean, clear-cut and mechanical', but he specifically distinguished this tendency from the admiration of machinery found in Marinetti, which he dismissed as a 'deification of the flux'.[21] Hulme's preferred art would tend towards fixity and that which is 'solid and durable'; formally, this art would 'culminate in a use of structural organisation akin to machinery'.[22] What Imagist poetry celebrated in the machinery of the Underground was not necessarily speed, but rather its ability to stage a poetic encounter which could stress fixity amid the vertiginous bustle of modernity.

The world's first underground was the London Metropolitan Railway, completed in 1863, running just below ground level from Paddington to Farringdon Street, and following the lines of existing streets.[23] In its second year the Metropolitan Line carried some 12 million passengers, and pioneered 'workmen's tickets', reduced fares on one early morning train for workers entering the city from the East End.[24] Such fares show how divisions within society were replicated in the spatial structures of transport. The first proper 'tube' railway opened in 1870, and the first electric tube appeared in 1890 with the opening of the City and South London Railway (now part of the Northern Line). From that date onwards a network of tubes grew beneath London, extending to a system 40 miles long by 1914.[25] The Underground promoted itself from early on as a convenient way to avoid surface traffic jams due to the burgeoning of horse-drawn tramcars from 1870 onwards (see figure 1).

The tube and other new transport systems, such as the electric tram, aided suburbanisation, moving the working class from inner-city slums to new suburbs like Islington and Clapham. These demographic changes often produced social conflict over the occupation of urban space. The extension of a tramcar line to Hampstead in 1882, for example, was opposed by residents on the grounds that it would lower the middle-class tone of the neighbourhood.

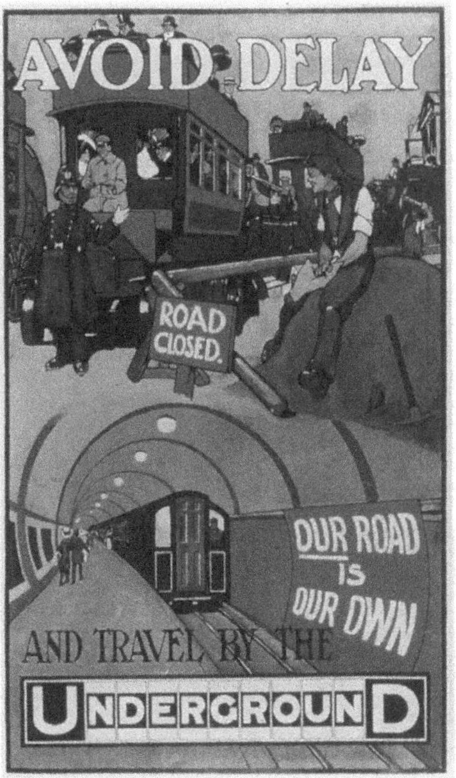

1 'Avoid delay', London Underground poster, 1910

Modernised transport prompted a profound restructuring of urban living space. The building of 'garden suburbs' and 'Metro-Land' housing estates, constructed in Buckingham, Middlesex and Hertfordshire with the co-operation of the Metropolitan Railway, produced the characteristic drift of the affluent classes to leafy environs removed from their places of work. Publicity for the 'Country Homes' building programme of the Metropolitan Railway Company in 1909 praised their tube trains for 'affording to City men a quick and frequent service of trains between the City itself and the various residential districts'.[26] Golder's Green, for example, was rural until 1907, when the tube opened there and urbanised the area, doubling the population by 1923.[27] Golder's Green (see figure 2) was promoted by the Underground as a place in which to escape 'the dull and smoky city', and was claimed to be 'London's healthiest suburb'.[28]

One commentator in 1910 noted that 'in this strenuous age business men find it necessary to spend much time in the Metropolis, and yet yearn

2 'Into the clean air and sweet sunshine (Golder's Green)', London Underground poster, 1911

for the charm, healthfulness and repose of a country residence'.[29] Although most of the early tubes were designed for such commuters, the Central London Railway (1900) provided a popular and inexpensive service, known as the 'twopenny tube', from west London for theatregoers and shoppers in Oxford Street (see figure 3).

In T. W. H. Crosland's splenetic assault upon suburban life *The Suburbans* (1905), the 'twopenny tube' is described as 'that triumph of modern suburbanism' and a 'ghastly burrow'.[30] The book has separate chapters on areas of London that are suburban, all of which – Clapham, Tooting, Kilburn, Hampstead and St John's Wood – possessed tube stations at this time. Crosland also devotes a chapter to methods of travelling to the suburbs, and offers a lengthy description of the horrors of travel by tube. The trains exhibit 'grime and heat and smell and rattle … beyond endurance', and as you stand up you are forced to inhale 'a mixture of tunnel air, condensing steam, and rank tobacco'.[31] Crosland's strongest

3 'For business or pleasure', London Underground poster, 1913

contempt, however, is reserved for the passengers. In the commuters there is, he notes, an 'absolute complacency and unthinking satisfaction' with the tube journey to the suburbs; there is 'not one man in a thousand who rebels and stands aghast at the whole business'.[32] The tube was thus, for the conservative Crosland, another instance of a modernising democratic spirit infusing the geographical restructuring of society.

This transport revolution was perhaps the main experience of technological modernisation for the vast majority of London citizens. Publicity for the extension of the line from Charing Cross in 1907 proclaimed that the Underground will 'introduce wide-ranging changes in the distribution of population, the location of shopping centres and the travel habits of people'.[33] The Underground thus played a key role in the reshaping of social space for millions of Londoners. Other technical innovations followed the tube: the first escalators began at Earl's Court in 1911 and electric lifts were common by 1907.[34] Early Underground posters stressed the modernity of the service, employing the slogan 'Light, power and speed' (see figure 4).[35] The first maps and a standardised design of station signs appeared between 1906 and 1908; the distinctive Underground typeface in 1916.[36] Encountering such phenomena profoundly altered the visual experience of space and time in the city. The proximity of certain places in the city becomes distorted by the time taken to traverse them. One's sense of the actual distance between, say, Waterloo and the Embankment,

4 'Light, power and speed', by Charles Sharland, London Underground poster, 1910

is governed by the time taken to travel under the river by tube. The representation of space on the tube map thus distorted the actual distance between places, altering one's sense of space and time in the city.

Phenomena such as these emphasised one of the experiences common to all rail travel: the shrinking of space and of the time taken to cover it. One earlier commentator on conventional railways commented: 'Distance is abolished, scratch that out of the catalogue of human evils.'[37] In 1841 Heinrich Heine claimed that, with rail transport, '[s]pace is killed by the railways, and we are left with time alone'.[38] As Wolfgang Schivelbusch notes, such extreme reactions are the effect of a profound disorientation in our 'perceptive powers', producing a feeling that space and time are being destroyed. In effect, the growth of railway transport heightened the desire for a new 'cognitive map' of modernity. One contributor to a medical congress of 1866 complained of the change in habits brought by travel: 'Conversation no longer takes place except among people who know each other, at least not beyond the exchange of mere generalities; any attempt to go beyond these often lapses due to the indifference of some travellers.' Loss of human conversation means that 'reading becomes a necessity'.[39]

Looking thus usurps talking or listening. This point, central to the experience of urban modernity, was stated succinctly by the sociologist Georg Simmel in 1903:

> Someone who sees without hearing is much more uneasy than someone who hears without seeing ... Interpersonal relationships in big cities are distinguished by a marked preponderance of the activity of the eye over the activity of the ear. The main reason for this is the public means of transportation. Before the development of buses, railroads, and trams in the nineteenth century, people had never been in a position of having to look at one another for long minutes or even hours without speaking to one another.[40]

Psychic unease at being situated in the 'giddy space' of the metropolis is intensified on tube journeys. On conventional trains – and on the few overground parts of the system – the stimulus to look can be alleviated by snapshots of the landscape through which one hurtles. But even that 'synthetic philosophy of the glance'[41] is denied to the Underground passenger, as objects for the gaze are drastically curtailed on the tube. The anxiety of the experience of urban modernity is increased: stimulated by the social processes of transport to gaze, one is prevented by the organisation of space within the train from viewing anything but other people. Early tube-trains on the City and South London Line were nicknamed 'padded-cells' because they lacked windows, and their low levels of lighting prevented reading. Lacking route diagrams, guards informed travellers of the stations, increasing one's spatial disorientation.[42] Crosland also grumpily noted this feature of the Underground: 'You rattle along through tunnels only just big enough to let your train through. At every station the attendant opens his glass doors, and rattles his iron wicket, and calls out the name of the station in more or less perfunctory accents.'[43] These early trains lacked advertisements, although that was soon to change, with adverts being placed over window spaces and then, with regulation, in the areas between and above windows.[44] Advertising worked in the tube because of the arrangement of visual space: at least now one could gaze at images without incurring the potential social embarrassment of exchanged glances with other passengers.

The tube is the most urban of transport systems, one which required travellers to relearn their perceptual relationship to the environment. By the end of the nineteenth century conventional over-ground rail transport was accepted, and journeys through countryside became, as they rarely are today, experiences of leisured visual consumption. However, in the early years of this century the tube required a new cognitive guide to parallel the geographical map of the Underground. Negotiation of city space via the tube is an ambivalent experience, especially for the Imagist poet determined to capture the essence of modern life. As Elizabeth Wilson comments, searching for meaning in the city takes various forms, 'not the least important of which is to create new forms of beauty'. However, such beauty 'will never be without a kind of unease'.[45] One must cope with the anxieties produced by encounters with the new spaces, and their specific visual regimes, but also celebrate the fresh experiences and pleasures offered. Initial reactions to spatial change often consist of mourning the loss of established sureties. But these changes also represent the myriad possibilities of modern metropolitan life, to be transformed and recorded as instances of aesthetic beauty.

Imagist excursions

Locked inside small tubular carriages, your gaze could fix on only two sorts of images: adverts or people. Imagist poems of tube travel show how urban *angst* was negotiated via visual images which then became aesthetic artefacts. It is hardly surprising that the first defiantly modernist group of Anglo-American poets should call themselves Imagists: the appellation merely drew upon the 'cult of images' pressing all around one in the spaces of modernisation.[46] By considering a number of poems by the male members of the Imagist movement – Richard Aldington, F. S. Flint, John Gould Fletcher and Ezra Pound – we can understand how the experience of 'time–space compression' described by Harvey is connected to a social and sexual 'logic of visualisation' in modernist space.[47]

The manner in which urban anxiety metamorphoses into the basis of a new aesthetic is evident in Richard Aldington's poem 'In the Tube', published first in *The Egoist* in 1915:

> The electric car jerks;
> I stumble on the slates of the floor,
> Fall into a leather seat
> And look up.
>
> A row of advertisements,
> A row of windows,
> Set in brown woodwork pitted with brass nails,

A row of hard faces,
Immobile,
In the swaying train,
Rush across the flickering background of fluted dingy tunnel;
A row of eyes,
Eyes of greed, of pitiful blankness, of plethoric complacency,
Immobile,
Gaze, stare at one point,
At my eyes.

Antagonism,
Disgust,
Immediate antipathy,
Cut my brain, as a sharp reed
Cuts a finger.

I surprise the same thought
In the brasslike eyes:

'What right have you to live?'[48]

Aldington casually displays what he called the Imagist method of having 'an eye for the common people, even if it be only to pity or to hate them'.[49] Both emotions are clearly evident. Aldington also revives Crosland's lost spirit of rebellion against the uniformity of tube travel. The final question can be read either as the protagonist's opinion of the 'row of eyes' of the fellow travellers, or as the antagonism of the 'eyes' of the poet. The image of the reed cutting the finger indicates the violence of perception between artist and people: this is a gaze of mutual antipathy. The poet, however, has the upper hand since he can try to represent and control the gaze confronting him. The Imagist drive to a concise language enacts a synecdoche of 'eyes' of 'pitiful blankness' to replace the other passengers. People are fragmented into component parts; these parts are then viciously solidified. Faces are 'hard', and eyes, using a term displaced from the 'brass nails' embedded in the train, become 'brasslike'. This reifying gaze signals not only contempt towards the non-artist, but also the defensive strategy of an aesthetic group threatened by the 'blankness' of these eyes. Concentration upon the fearful gaze of the public, however, springs initially from the visual space inside the tube. Aldington's fixing of human subjects originates in the 'row of advertisements' upon which his gaze first falls. The ordered structure of this 'row' gets displaced onto other signifiers: the row of advertisements becomes a row of windows, which becomes a row of faces, only to conclude as a row of eyes.

The travellers in the poem are not only punitively anatomised by the poet's gaze, but are rendered static and 'immobile'. The whole text vacillates between movement and *stasis*, mirroring the spatio-temporal experience of travel. The language of the poem itself jumps from condensed single-word lines ('Antagonism,/ Disgust') to longer, jerkily moving, lines such as 'Eyes of greed, of pitiful blankness, of plethoric complacency'. Faces are 'immobile' but are situated within the 'swaying train', travelling at speed through the 'dingy tunnel'. Aldington's gaze seeks to arrest antithetical faces, preventing them realising their violent dislike of him. Ferocious gazing at antagonistic eyes will, the poet hopes, contain their threat, transforming their active staring into the fixity of the brass nails embedded in the woodwork of the train. Eyes that were 'pitiful' would thus become 'pitted' like nails, their pointed ends safely hidden from view. The eyes and their attendant bodies become images safely affixed to the train. Pumped around at velocity within the metropolis, people become like adverts, images proclaiming their availability for visual consumption.

F. S. Flint's poem 'Tube' presents a similar experience of travelling. Here the travelling public is 'stolid', and sit

> Lulled
> By the roar of the train in the tube
> Content with the electric light
> Assured, comfortable, warm.[50]

Once more the focus is upon the eyes of other travellers: 'You look in vain for a sign,/ For a light in their eyes. No!' In contrast to Aldington's text, where active public eyes must be restricted, Flint's poem represents a desire to energise the public: 'And we, the spirit that moves./ We leaven the mass,/ And it changes.' The public must be propelled into motion out of their inertia, infused, as poets are, with 'the spirit that moves'. In Flint's text this artistic 'spirit' derives from the current powering the tube and its 'electric light', and is commandeered by the poet to disturb the 'mass, inert,/ Unalarmed, undisturbed'. Technological modernisation, altering the experience of space and time, here informs the very textuality of writing.[51] Only poets can 'sweeten' and 'leaven', engineering a modernising rejuvenation of the public, otherwise 'the world/ Would sink in the ether'. Flint is anxious to distinguish the modernist poet from the anonymous 'mass' of the other travellers, to carve out a social distance from them that is not evident in the space of the tube train. Travel by tube produces the perilous necessity to try to individualise one's identity and thus distance oneself from the lumpen mass. Traffic in urban environments forces a mingling in space of social, sexual and class relations. The desire to 'leaven' the mass was thus prompted by the poet's encounter with large numbers of people while travelling.

Echoing Ford's advice to young poets, T. E. Hulme remarked: 'The beauty of London [is] only seen in detached and careful moments', such as when the artist is 'in some manner detached', travelling 'on top of [a] bus'.[52] Representations of meetings while travelling provided new everyday material for modernist verse and helped indicate a modernity of subject matter for the poet. But such liaisons also allowed the writer to remain aloof from the 'mass' by means of the safe detachment of the gaze. This distance helped uphold the poet's superior status in a world where the economic market was ruthlessly equalising old social hierarchies.[53] The poems by Aldington and Flint demonstrate how they valued the social distance and reserve that the spaces of transport offered. In 1902 Simmel argued that the individual's sense of reserve in crowded cities was 'because the bodily proximity and narrowness of space makes the mental distance only the more visible'.[54] The experience of the 'narrowness' of the London Underground, in Aldington's case, transforms 'reserve' into 'antagonism'.

This conflict between the individuality of the artist and the mass travelling public resounds throughout the modernism of this time. It is part of what Clive Bloom identifies as the defining preoccupation of literature in Britain in the years from 1900 to 1929: 'the tension between the individual and the collective (the state, the class, the gender, the union, the race)'.[55] In the case of the Imagists, individualism was linked to a semi-philosophical or political position articulated in the pages of *The Egoist: An Individualist Review*, the magazine that did much to promote the movement and on which Pound, H. D. and Aldington all worked.[56] Individualism was also, as Robert von Hallberg notes, a way of distinguishing Imagism from the collective identity of Futurism, and a way of reconciling a group position with the artistic conscience of an individual poet.[57] The 1916 anthology proudly proclaimed of the Imagist that 'they are Individualists'.[58]

The main composition of the mass confronted by the individualist Imagist would probably have been clerks employed in the vastly expanded City of London, although their income would have been above that of Forster's Leonard Bast who walks rather than take buses.[59] Richards and MacKenzie note:

> The inhabitants of the new suburbs were in the main middle class. They were drawn from that army of white-collar workers, managers, administrators, lawyers, technicians, and clerks which had been called into

> existence by Britain's commercial and financial pre-eminence. It was these men who travelled into London each day by train from Clapham, Wimbledon, Richmond, Putney, and Barnes.[60]

For the working class, the cost of regular tube-travel was prohibitive, and certain railway companies, eager to keep their suburban customers happy in splendid isolation, resisted attempts to encourage workers to use trains.[61] It was thus this class of suburban travellers that provoked the antagonism of the Imagists, and which affronted their sense of individualism.

One of the more sustained Imagist texts on urban transport to explore this theme was John Gould Fletcher's 'London Excursion', first published in *The Egoist* in 1914 and then reprinted in the Imagist anthology of 1915. The poem finds the protagonist forced into intimate spatial relations with the commuting masses from the new suburbs ringing the city.[62] Fletcher's poem is in nine sections, and recounts travel, initially by bus, across the capital from morning until early evening. Paradoxically, at this time, in 1914, Fletcher was himself living in Sydenham, a Kent suburb, and travelling into London daily to work, not as a clerk, but as an aspiring modernist poet. 'London Excursion' was written, noted Fletcher, to 'commemorate my journeyings to and fro, between the domestic half of my existence at Sydenham and the literary half in London'.[63] His journey on the 'red bulk' of the bus through the 'angular city' summons the familiar feeling of urban alienation, and is also structured around flux and fixity:

> Passivity,
> Gravity,
> Are changed into hesitation, clanking pistons and wheels.
> The trams come whooping up one by one,
> Yellow pulse-beats spreading through darkness.
>
> Music hall posters squall out:
> The passengers shrink together,
> I enter indelicately into all their souls.[64]

Though the passengers 'shrink' together for comfort the poet seeks to isolate himself from real connection with them: 'Yet I revolt: I bend, I twist myself,/ I crawl into a million convolutions./ Pink shapes without angle'. This desire to escape the traffic of the city is thwarted, however, when a 'long hot bar' from the bus 'pierces the small of my back'. In a section entitled 'Bus-top' the poet looks down and sees 'Monotonous domes of bowler-hats/ Vibrate in the heat'. The uniformity of the urban commercial worker, recalling the 'rows' in Aldington's poem, is noted from a position which marks the poet's spatial distance above the object of his gaze: his visual skills separate him from 'monotony'. Amid this detached vision the poet has a 'Sudden desire for something changeless', for something '[u]nmelted by hissing wheels'. This is a desire for some firm grasp, a discourse of place, among the 'melting' modernising city with its '[n]oise, uproar, movement'. But the protagonist is unable to locate himself: 'I can no longer find a place for myself:/ I go.' He is 'blown like a leaf/ Hither and thither' amid the 'gyration' of the city, and has no choice but to join the '[s]traggling shapes' of commuters about to depart for home: 'A clock with quivering hands/ Leaps to the trajectory-angle of our departure.' The revolt against urban life fails, and the poet finds himself controlled by that powerful system of modernist time, the station clock. No 'essence' of urban life has been discovered, and the poet travels home to the 'green' suburbs with the wearers of the bowler-hats he had earlier perused with contempt. A similarly resigned tone is discerned in a slighly later poem by Fletcher, 'From Babel's Night', set amid the 'great dark cave of steel' of a tube station. Here, waiting for the last train at night, are

> the wrecks, the lost,
> Who pass out, drugged with weariness,
> From dark illusion's shrine
> To suburbs in the night.'[65]

'London Excursion' marks a failure, the inability of the Imagist poet to maintain a necessary 'distance' from the people despised, or to fix his own identity in the swirling spaces of the city. F. S. Flint's 'Accident', from the 1915 Imagist anthology, shows gender and not class as the locus of the personal relations imposed on urban space, and shows fixity to be directed at another passenger:

> Dear one!
> You sit there
> in the corner of the carriage;
> and you do not know me;
> and your eyes forbid.[66]

In the 1930 Imagist anthology, Richard Aldington wrote that 'what enters by the eyes is desire', and when desire

speeds through the eyes
For a moment there is strange tumult
In the whole nature of man or woman,
And in a flash all life is changed.[67]

In Flint's poem the moment of transformative desire is prompted by the chance visual encounter with a stranger whose 'eyes forbid' further intimacies. The woman can 'see beyond' and understand that the other passengers are 'nothing'. This visionary gift propels the woman to transcend the 'dead faces' and 'wear of human bodies'. The poet desires this ability in order to project himself out of the 'dirt' and 'squalor' of the urban context, symbolised by the faces of these neighbours. Watching, the woman turns swiftly from 'love' to 'desire', but this gaze is curtailed by the train arriving at the protagonist's destination: 'This is my station ...'. The poem shows perceptual desire, 'tense and tender', thwarted by the necessity of leaving the train. The 'accident' of the encounter with the 'dear one' is constrained by the systematic nature of the travelling: one has to leave at one's destination and make do with creating a poem from this experience of desire within intimate space.

Women in the 'closing years of the nineteenth century', comments Wilson, 'were emerging more and more into the public spaces of the city'.[68] But their presence on public transport were often organised so as to minimise the sort of accidental meeting with the other sex portrayed in Flint's poem. Following American practice, 'Ladies Only' carriages appeared on British railways, along with separate waiting-rooms for men and women on stations.[69] In nineteenth-century Paris women were forbidden to travel on top of buses, preventing the visual pleasures advocated by Hulme and Ford.[70] Smoking-carriages were assumed to be for men only, as is shown in Virginia Woolf's novel *Jacob's Room* (1922); Jacob enters a carriage occupied by a solitary woman, Mrs Norman, who tells him nervously: 'This is not a smoking-carriage.'[71] The early 'padded cell' tubes contained one car for smokers from which women were banned.[72] In 1910 *The Railway and Travel Monthly* discussed the growing number of women smokers, quoting the advice of *Punch:*

Here's the place where men may smoke;
Not designed for women-folk.
If they come in solid packs,
Take and put them on the racks;
Should they faint or weep or shout;
Open the door and drop them out![73]

If there is evidence that the volume of women travelling increased by the end of the nineteenth century, their visibility was still subject, not only to the 'male gaze', but to spatial and social organisation. It was only with the 1914 War that women began working in any numbers on the railways.[74] Women were first recruited to the Underground in March 1915, and in that June Maida Vale Station boasted an all-female staff. As Sandra M. Gilbert has argued, these uniformed women challenged the male modernist occupancy of the metropolis.[75] Women also contested city space in the marches and spectacular demonstrations mounted by the suffragettes in London between 1906 and 1913.[76] The 1908 suffrage protesters, for example, were described as 'an innumerable swarm of humanity': the crowd was 'a flood with its slow but steady currents, setting hither and thither'.[77] Such city crowds, as Wilson notes, became perceived as essentially 'feminine' and figured as 'hysterical, or, in images of feminine instability and sexuality, as a flood or swamp'.[78] Wyndham Lewis, for example, viewing the pre-war London crowds in July 1914, pondered, 'Are the Crowds then female?', referring to them as a 'feminine entity' of 'meaningless numbers'.[79] Such public displays of femininity in the city compromised the detached male pleasure in looking, desired by Imagism.

Ezra Pound's famous definition of the 'Image' was that it presented 'an intellectual and emotional complex in an instant of time'; its effect was 'that sense of sudden liberation; that sense of freedom from time limits and space limits ... which we experience in the presence of the greatest works of art'.[80] Perhaps it is not too fanciful to view this desire for aesthetic transcendence as conditioned by the crises in perception wrought by the 'time–space compression' of modernity, central to which is a sense of the failure of the aloof male gaze. Elizabeth Wilson argues that the *flâneur* 'embodies the Oedipal under threat. The male gaze has failed to annihilate the castrate, woman. On the contrary anonymity annihilates *him*.'[81] This scenario is exacerbated when the stroller becomes the traveller. In Flint's 'Tube' we see the poet's desire to distinguish himself from the anonymity of the 'mass', an experience shared by the protagonist of Fletcher's urban excursion; Aldington's journey shows his gaze countered by those of other travellers. The most famous of Imagist poems, Ezra Pound's 'In a Station of the Metro' shows the poet striving to manufacture beauty from the mundane surroundings of the Paris Metro. The poem also registers the desire to escape the suffocations of the crowd by producing a distance sufficient to view with detachment the faces of other passengers. I want to look at this poem in some detail, to demonstrate the importance of a spatialised reading of Imagism. By examining Pound's account of the poem's composition in his essay 'Vorticism' we can see how the text displays clearly the beauty and unease of modern city spaces for the masculine traveller.

Faces in the Metro

Pound's account of the composition of 'In a Station of the Metro' in 'Vorticism' attaches special emphasis to his vision of anonymous faces: 'Three years ago in Paris I got out of a "metro" train at La Concorde, and saw suddenly a beautiful face, and then another and another, and then a beautiful child's face, and then another beautiful woman.'[82] The stress on the suddenness of the incident accords with the 'sudden liberation' of the Imagist complex. This sense of the instant is matched by the set of separated faces Pound describes; these are particular faces that Pound seems unwilling to combine, the cumulative syntax of 'and then' and 'another' stressing their discreteness.

The continuation of Pound's narrative shows that his initial difficulty in finding words to capture his experience ('I could not find words that seemed to me worthy, or as lovely as that sudden emotion') eased when he turned to a visual discourse. Walking home that evening, a form of expression for the emotion came to him, not actual words but 'an equation ... in little splotches of colour'. It was, notes Pound, the start of 'a language of colour'.[83] It is strange, then, that very little colour appears in the finished text. Instead, the image is drained of colourful intensity, surrendering to the bleak environment of the Metro platform. As quoted in 'Vorticism', the poem appears as follows:

> 'In a Station of the Metro'
>
> The apparition of these faces in the crowd:
> Petals on a wet, black bough.[84]

Finding beauty in the urban crowd is a difficult experience for Pound, as for the other Imagists. Beauty in the Metro lies not in the crowd, not in a person in the crowd, but only in their faces. However, this beauty is only evident at a perceptual distance and is devoid of Pound's initial 'language of colour'. In his other account of the composition of the poem, Pound refers to being jostled by the departing passengers.[85] Touch is thus replaced in the poem by the detachment of the gaze. For it is not the faces-in-themselves that are beautiful, but their 'apparition' as 'petals' which is pleasing to the poet. As Maud Ellmann notes, 'the text records the "apparition" rather than the faces per se'.[86] Pound's claim that the poem is of 'a thing outward and objective' is vanquished by the subjectivism of the perceiving consciousness of the apparition. The poet does not just record objective 'arrangements in colour'[87] but displays his emotions before the objects. It is a feeling which, as in Aldington, an-atomises human beings into mere 'faces'. Pound's 'faces in the crowd' are merely fixed images to be consumed, 'petals' to be plucked by the gazing poet as objects of beauty.

The poem matches the negative emotions of other Imagist texts before crowds, but in a more aestheticised and disguised manner. This threat to individuality was also diagnosed, in terms strikingly similar to Pound's, by C. F. G. Masterman in 1909:

> It is in the city Crowd, where the traits of individual distinction have become merged in the aggregate, and the impression (from a distance) is of little white blobs of faces borne upon little black twisted or misshapen bodies, that the scorn of the philosopher for the mob, the cynic for humanity, becomes for the first time intelligible.[88]

Crowds blur the particularity of individuals, producing facial 'splotches'. The problem with these 'splotches' is that they threaten the spatial clarity of the textual image, spilling image into surrounding image.

Pound's solution to this visual quandary is found in the typography of the poem. It marks a small, but radical, break in the formal treatment of poetic language, an innovative representational space in Lefebvre's terms, that was to be developed in Pound's later writings. When originally published in *Poetry* in April 1913, this is the way the text appeared:

'In a Station of the Metro'

The apparition   of these faces   in the crowd:
Petals   on a wet, black   bough.

The poem's representational space is designed to resist sequence and time in favour of spatial arrangements. The poem not only fragments people into faces, but atomises its own language. The collectivism of the crowd is analysed into a series of linguistic monads. Language is literally chopped into separated components, each unit of discourse striving to render only one thing – the faces, the crowd, the petals, the bough. The words 'these' and 'the' indicate a definiteness of linguistic reference: it is not a crowd but *this* crowd, this word referring to this thing. Any sense of movement – as in the transportation process of the Metro itself – is resisted by the spatial form of the words. The text decomposes not so much into two complex images – faces and petals – which are then compared, but rather divides into six discrete images, six linguistic units parcelled off from one another by blank spaces. These six pictures now resemble nothing more than the rows of advertisements found in the Underground or Metro station, or within the train, and noted in Aldington's poem 'In the Tube'. The way adverts for different commodities are juxtaposed without any sense of their logical, narrative or economic connection is just how Pound's six linguistic things appear. Syntactical links are forced asunder by the Mallarméan blanks between the words, just as in Pound's prose account the faces were split by phrases such as 'and then' and 'another'. Now textual space, in a form of negative heterotopia, parallels other spaces – that between poet and passengers, and the gap between each of the 'faces'. The 'apparition' is not of Pound entering any relationship with these subjects. Instead he sees 'faces', as restricted in their movements as 'petals' pinned on the 'bough'.

Subsequent printings of the poem erased the spaces between the linguistic units. One explanation for this is Pound's advocacy of Vorticism as a 'dynamic' art. A poem syntactically falling apart from within is a poor example of the vortex, 'from which, and through which, and into which, ideas are constantly rushing'.[89] Pound later claimed that one problem with early Imagist poetry was that it 'thought only of the STATIONARY image'.[90] Pound's original poem had, almost too clearly, grasped the uneasy nature of the modern experience of space: atomistic, reified, and where the only obtainable sensuous satisfaction was that of a naked staring at 'beautiful faces'. And we are not even compensated by the promised myriad of pleasurable colours.

After quoting the poem in the essay 'Vorticism', Pound asserts that this kind of poem is an attempt to 'record the precise instant when a thing outward and objective transforms itself, or darts into a thing inward and subjective'.[91] The type of subjective emotion expressed by 'In a Station of the Metro' can best be illustrated by considering what might be seen as its companion poem from the volume *Lustra* (1916), 'Dans un Omnibus de Londres'.[92] An odd transfer of discourses occurs between this poem and that of the 'Metro' text. The poem about London travelling is written in French, while the poem about a French journey is composed in English. It is as if Pound requires the metaphorical distance of a foreign language in order to maintain his own aesthetic distance from the urban environment. The French text, however, clarifies the move from objective to subjective that defines Imagist perception:

Les yeux d'une morte
M'ont salûe,
Enchassés dans un visage stupide
Donts tous les autres traits étaient banals,
Ils m'ont salûe
Et alors je vis bien des choses
Au dedans de ma mémoire
Remuer,
S'éveiller.

(The eyes of a dead woman
Greeted me,
Set in a stupid face
All the other features of which were ordinary,
They greeted me
And then I saw many things
Inside my memory
Move,
Awaken.)[93]

Closer in tone to Flint's poem 'Tube', this objective image of looking on a London bus prompts a feeling of the woman's mundane nature. Pound is able to shift from this object to his 'mémoire', his subjective inward emotions. The lines which follow detail images of ducks, pigeons and 'deux petite filles graciles' seen in the Parc Monceau, Paris. English urban space, with its contemptible female travellers, is replaced by the space of memory, pleasingly French and pastoral. Later the poem recounts hiring chairs and watching

les cygnes noirs,
Japonais,
Leurs ailes
Teintées de couleur sang-de-dragon,
En toutes les fleurs
D'Amenoville.

This is much more clearly a poem employing a language of colour, compensating for the supposedly 'visage stupide' the poet perceived on the bus. The beauty Pound found in the faces at La Concorde Metro is here sadly missing; he has to recreate colourful female images to escape the dead eyes of the woman who has, tellingly, confronted him. The poem's subjective impression thus retreats from fidelity to the perceived object. Urban travel stimulates the desire to look, but when the object viewed is deemed too 'ordinary', then Pound's own discourse travels from London to Paris, trying to recapture the beautiful objects of an earlier set of perceptions. Dead faces of 'stupid' women could be countered by images of beautiful women. In 'The Condolence' Pound wrote: 'Our maleness lifts us out of the ruck.'[94] The ruck, that is, of the crowd that smothers particularity and the detached male voyeur. Trying to be hoisted above the social space of the city crowd here involves the male gaze desiring desperately to create beauty from the everyday.

Zestful images

Visual transcendence, the creation of the Imagist 'image', is an aestheticised attempt to escape the social and sexual constrictions of urban 'time–space compression'. But, as Pound's poems show, the experience of urban space is too strong to be simply jettisoned. Social space forces itself in the 'Metro' poem into the very crevices of the text. In this way we witness how the culture of transport affects the spatial form of Imagist poetry: the textual space of Imagism is thus constituted by this encounter with the complex social space of the city. Urban transport also creates conditions which privilege the visual sense. But the worrisome space of that experience, where sight is isolated from other sensual human relations, entails either a desire to transcend the situation – as in Flint's 'Accident' or Pound's 'Metro' – or an antagonism towards those upon whom one's gaze falls. Often transcendence merges into a violent visual gaze, as in Aldington's 'In the Tube'. Both are responses to the socio-spatial relations of the city. By fixing others in an aesthetic image urban anomie can be confronted.

This strategy influences Imagist texts that are not specifically about urban space. Richard Aldington, discussing how to write an Imagist poem describing a beautiful woman, notes that instead of utilising wasteful adjectives, 'we present that woman, we make an "image" of her'. Such an Imagist poem would possess 'hardness, as of cut stone. No slop, no sentimentality', and might be 'nicely-carved marble'.[95] This desire for concretised concision, manifesting itself in the characteristic shortness of the Imagist poem, is stylistic advice drawn from the urban environment, where brief glances are the norm.[96] Instilling such snapshots with intense significance and making them solid, like marble, grants a permanence denied to the glimpse of a stranger's face on the tube. Drawing on Bergson, T. E. Hulme theorised poetry as a 'visual concrete' language, a 'compromise for a language of intuition which would hand over sensations bodily. It always endeavours to arrest you, and to make you continuously see a physical thing.' Hulme then provides an example of how this visual reification operates: 'If you are walking behind a woman in the street, you notice the curious way in which the skirt rebounds from her heels.' If, notes Hulme, this 'motion' interest you, then you will search for the 'exact epithet' to convey this 'aesthetic emotion'. Hulme concludes that 'it is the zest with which you look at the thing which decides you to make the effort'.[97]

Aesthetically arresting, the 'motion' of the woman produces a visual concrete image. If poetry is a compromise for another discourse, one which enacts bodily sensations, then we see here the sexualised nature of Hulme's theory. The transcendent 'zest' of the gaze prompts the poem, the bodily desire which must be diverted into visual pleasures that produce the correct aesthetic. Hulme's scopophilic example also intrigues because the voyeur is *behind* the woman in the street, spatially safe from being gazed at in return. One of Hulme's poetic 'images' also illustrates this idea: 'Her skirt lifted as a dark mist/ From the columns of amethyst.'[98] The textual space of the 'zestful' image is, then, informed by the response of the male poet to the urban experience. The metaphor of solidity for the woman's body ('columns of amethyst') halts her movement, and thus eases the poem out of the limits of time and space: it is a case of all that is moving is fixed into stone. The aesthetic success of the poem, the production of a concrete and striking image, relies upon reifying the subject of the poet's glance. In Hulme's prose works a similar structure is found. 'Words', writes Hulme in 'Notes on language and style', are 'seen as physical things ... Want to make them *stand up* ... e.g. walking on a dark boulevard. Girl hidden in trees passes on other side. How to get this.'[99] Again the aim is to capture an experience of looking at a woman, but the setting of this scenario is more sinister. Hulme requires a language able, slightly comically, to 'stand up'. In another moment of linguistic tumescence linked to looking, he writes: 'A man cannot write without seeing at the same time a visual signification before his eyes. It is this image which precedes the writing and makes it firm.' Hulme, perhaps wistfully, concludes this point by saying: 'Solidity [is] a pleasure'.[100] This pleasure in the solid is, therefore, linked to the creation of the visual image in poetry. Hulme's criterion for successful poetry is thus: 'Is there any zest in it? Did the poet have an actually realised visual object before him in which he delighted?'[101] Realising an object, another version of Pound's notion of the transcendent image, thus relies upon reifying the female form as an object for male contemplative 'zest'.

We have seen how the transcendent 'image', rooted in a version of the male gaze, is a gendered response to the spatial and social conditions of the modernist city. Discerning the politics of the gaze, as represented in a cluster of Imagist poems, relies upon untangling the psychic and social relations between looking and the politics of space. In these poems textual space is infused by the social spaces of modernism. But it is important to understand how all social space, as Lefebvre argues, is composed of a variety of forms, here simultaneously psychic space and sexual space. Urban social space produces the conditions necessary, but not sufficient, for the mastering 'zest' of the gaze. Certain forms of transport in urban cities are one important example. This is not to suggest that forms of travel inherently enforce male visual power. For the male Imagist gaze has only a tenuous grip on these spaces: fixing others in images, or rooting oneself apart from others, is the course the Imagists took to subdue the terrors of urban disorientation. Instead of seeking comfort in the representation of place – as in *Howards End* – Imagism sought to reduce the threatening force of the flux of modernity by reifying images of others (or of the self) in their representational spaces.

It seems appropriate, as in many tube journeys, to end where we started, with a contemporary version of the textual space of the city. By this means we can evaluate the distance we have travelled from Imagist poems about the early Underground. Jonathan Raban's *Soft City* (1974), used by David Harvey as a symptomatic account of postmodernist city life, contains descriptions of traffic which have their ancestors in the Imagist poems I have been considering: 'Back among strangers – the essential condition of metropolitan life – straphanging in a rush hour tube, everyone avoids each other's eyes.' For Raban 'city-life' consists of 'belonging' mixed with anonymity, 'when one is stripped of credentials and credibility, when one sees oneself as just another moon-face in the crowd'.[102] Urban alienation – and the preponderance of the visual – mark the experience of this space as essentially unchanged since modernism. The desire to 'belong' is ruthlessly denied by the speed at which one travels, preventing one becoming rooted in any one space. Absorption into the 'faces' of the crowd is still a threat, as it was for Pound, Flint and Fletcher. Raban continues by indicating the sexualisation of looking in such situations: 'In the crowd, I catch the eyes of girls – that rapid, casual, essentially urban interrogation: a glance held between a man and woman fixes each for an instant; another belonging, of a kind.' The lesson of such encounters is that a 'big city is an encyclopaedia of sexual possibility, and the eye language of the crowd asserts this possibility without the risk of real encounters'.[103] These possibilities, notes Raban, are 'best at a distance – when you are on an up-escalator while the girl is travelling down; through carriage windows; across streets thick with traffic'. They are not 'invitations to an assignation' but merely 'the small assuring services which men and women can render for each other' when 'sucked in and reduced by the crowd'. That the gazing which Raban notes is more reciprocal than in the Imagist poets discussed above indicates also the changing social conditions surrounding gender relations across the century.

The momentary glance of Imagism appears once more, this time viewed both as detached sexual possibility and as 'reassuring services' for those riddled with urban *angst*. People are brought together in the spaces of transport, but intimacy is replaced by the less risky action of gazing 'at a distance'. For Raban the sexual gaze is a way of overcoming the anxiety of being 'reduced' by the city to just another featureless face. If one cannot 'belong' to this urban place, then one can at least transcend it via an imagined sexual encounter. Gazing is a reciprocal relation, an Hegelian move of mutual recognition, in which both sexes attempt to understand and 'belong'. Are Raban's 'reassuring services' another reformulation of the male gaze, assuaging a troubled masculinity by the 'interrogation' of women representing 'sexual possibility'? Or is this episode mournfully heroic, what Elizabeth Wilson describes as the attempt of both sexes to survive in the 'disorientating space … of the metropolis' by finding meaning, beauty and individual identity wherever possible?[104] Whichever interpretation we choose it is clear that the spaces and images of modernism are still ours to contest and revise.

This discussion of Imagism has shown how two aspects of Lefebvre's hypercomplex social space are deeply intertwined: the geography of technical modernisation shown in the tube transport, and the psychic spaces of an urban 'logic of visualisation'. The two forms of space cannot, in practice, be separated and are central to the textual space of Imagism. The uneasy beauty of Imagist poetry thus condenses together social, psychic and textual space, showing how relations of power, as Foucault argued, always infuse spatial relations. In another sense, we might view the Imagist poem set on the tube as a kind of negatively experienced version of Foucault's heterotopia: a place without a place, but located in material reality, and structured by relations between sites (the stations), the experience of the Underground is composed of complex and contradictory relations between social, sexual and aesthetic spaces (OS, p. 24). For the Imagist poet the tube is a kind of crisis heterotopia, as Foucault terms it, one in which the poet stages an encounter with an other that he or she hopes will result in some place of fixity amid the incessant movements and spatial disorientations of modernity. Chapter 4, on Joyce's *Ulysses*, considers a rather more welcoming attitude to the experience of moving through a different city, that of Dublin.

Notes

1 Ford Madox Hueffer (Ford), 'Modern poetry', *The Thrush* 1: 1 (1909), 49. It is worth noting how certain Imagist poets seem to have taken Ford's advice quite literally: see, for example, John Gould Fletcher's long poem 'London Excursion', with its section 'Bus-top', which is discussed below.

2 Ford, 'Modern poetry', p. 49. The request to shift aesthetic attention eastwards is quite significant, given the way that this area of London was demonised as 'the abyss'. Ford had earlier analysed the different representational spaces of the east and west ends of London in his 1905 *The Soul of London: A Survey of a Modern City* (London, Everyman, 1995), p. 48.

3 T. S. Eliot, for example, stated: 'The *point de rèpere* usually and conveniently taken as the starting point of modern poetry is the group denominated "imagists" in London about 1910.' See Eliot's 'American literature and the American language' in *To Criticize the Critic* (London, Faber, 1965); quoted in Peter Jones, ed., *Imagist Poetry* (Harmondsworth, Penguin, 1972), p. 14. Most subsequent critics have taken this account at face value; see, for instance, the influential account given by C. K. Stead, *The New Poetic: Yeats to Eliot* (Harmondsworth, Penguin, 1967), ch. 4. Rainer Emig's more recent account, though it judges Imagism a limited success, does note the significance for modernism of its 'anti-poetic poetics' which tried to force language to be radically referential: Rainer Emig, *Modernism in Poetry: Motivations, Structures and Limits* (London, Longman, 1995), pp. 106–10. For a brief argument that Imagism is not the first avant-garde movement in modernism but its contrary see Lawrence Rainey, *Institutions of Modernism: Literary Elites and Public Culture* (New Haven, CT, and London, Yale University Press, 1998), pp. 29–30.

4 T. E. Hulme, 'The Embankment (The fantasia of a fallen gentleman on a cold, bitter night)'; taken from the manuscript version of the poem in A. R. Jones, *The Life and Opinions of Thomas Ernest Hulme* (London, Gollancz, 1960), p. 159.

5 Freud's technical definition of 'fetishism' fits Hulme's image well. See Freud, 'Fetishism' (1927) and 'Splitting of the ego in the defensive process' (1938) in *Collected Papers*, vol. 5, ed. James Strachey (New York, Basic Books, 1959). For an application of this concept to the visual arts see Laura Mulvey, 'Fears, fantasies and the male unconscious' in *Visual and Other Pleasures* (London, Macmillan, 1989).

6 Fredric Jameson, 'Postmodernism, or the cultural logic of late capitalism', *New Left Review* 146 (July–August 1984), p. 80.

7 See Susan Buck-Morss, 'The *flâneur*, the sandwichman and the whore: the politics of loitering', *New German Critique* 39 (1986), 99–140; Griselda Pollock, 'Modernity and the spaces of femininity' in *Vision and Difference: Femininity, Feminism and Histories of Art* (London and New York, Routledge, 1988); Elizabeth Wilson, 'The invisible *flâneur*', *New Left Review* 191 (January–February 1992), 90–110; and Janet Wolff, 'The invisible *flâneuse*: women and the literature of modernity', in Andrew Benjamin, ed., *The Problems of Modernity: Adorno and Benjamin* (London, Routledge, 1989). For a critical overview of these debates in relation to women's writing in the twentieth century see Deborah L. Parsons, *Streetwalking the Metropolis: Women, the City and Modernity* (Oxford, Oxford University Press, 2000).

8 Pollock, 'Modernity and the spaces of femininity', p. 67. The originator of the psychoanalytic theory of 'the gaze' is Laura Mulvey, 'Visual pleasure and narrative cinema', reprinted in her *Visual and Other Pleasures*. Mary Ann Doane argues that space is a key component of 'the gaze'. For Doane the structure of the gaze is not so much Mulvey's one of active:male–passive:female, as a model of proximity–distance in relation to the image. The (male) voyeur must maintain a distance between himself and the female image; this space is the lack which fuels desire. 'Spatial proximity' is seen as 'a female' attribute, to be resisted if the detached male gaze is to operate. See Mary Ann Doane, 'Film and the masquerade: theorizing the female spectator', *Screen* 23: 3–4 (September–October 1982), 77–80. For a cultural geographers' development of the concept of the gaze that uses the work of Judith Butler see David Bell, Jon Binnie, Julia Cream and Gill Valentine, 'All hyped up and no place to go', *Gender, Place and Culture* 1 (1994), 31–47.

9 Pollock, 'Modernity and the spaces of femininity', p. 71.

10 Pollock, 'Modernity and the spaces of femininity', p. 71.

11 See Walter Benjamin, *Charles Baudelaire: A Lyric Poet in the Era of High Capitalism*, trans. Harry Zohn (London, Verso, 1983), pp. 149 and 151.

12 E. M. Forster, diary entry, 21 October 1909, cited in Forster, *Howards End* (Harmondsworth, Penguin Books, 1968), p. 351.

13 Wilson, 'The invisible *flâneur*', p. 109.

14 Wilson, 'The invisible *flâneur*', p. 110.

15 Ezra Pound, 'Provincialism the enemy', *Selected Prose 1909–1965*, ed. William Cookson (London, Faber, 1973), p. 169. The tunnel envisaged here is one beneath the English Channel to connect London and Paris as cultural capitals of the world.

16 'Preface' to *Some Imagist Poets: An Anthology* (Boston and New York, Houghton Mifflin, 1915), p. vii.

17 Buck-Morss, 'The *flâneur*, the sandwichman and the whore', p. 102.

18 See Marshall Berman, *All that Is Solid Melts into Air: The Experience of Modernity* (London, Verso, 1983), pp. 15–16.

19 John Davidson, *Fleet Street and Other Poems* (London, Grant Richards, 1909), pp. 100–10. See also Davidson's early poem 'Song of a Train', where the train is a 'monster' that '[s]peeds through the land', in *Ballads and Songs* (London, John Lane, 1894), pp. 103–6.

20 Interview with Marinetti, *The Evening News* (London), 4 March 1912, p. 3. On Marinetti in London see Richard Cork, *Vorticism and Abstract Art in the First Machine Age* (London and Bedford, Gordon Fraser Gallery, 1975), vol. 1, p. 28. Peter Nicholls argues that Futurism's uncritical praise of technical modernisation distinguished them from the Imagists. Imagism and Vorticism, he suggests, subordinated 'the technological to the aesthetic' so as to uphold art against commerce. Both Futurism and Imagism were interested in space, but for Imagism 'space was redefined as distance, as a praxis of visual perception which allowed detachment and contemplation rather than immersion in the flows of capital'. See Peter Nicholls, 'Futurism, gender, and theories of postmodernity', *Textual Practice* 3: 2 (summer 1989), 210–11.

21 T. E. Hulme, 'Modern art and its philosophy', in Herbert Read, ed., *Speculations: Essays on Humanism and the Philosophy of Art* (London and New York, Routledge & Kegan Paul, 1987), pp. 97 and 94.

22 Hulme, *Speculations*, pp. 100 and 98.

23 John R Day, *The Story of London's Underground* (London, London Transport, 1963), p. 4. Unattributed information concerning the tube comes from the London Transport Museum, Covent Garden. For a stimulating account of the impact of the London Underground, especially the work of Frank Pick, see Michael T. Saler, *The Avant-Garde in Interwar England: Medieval Modernism and the London Underground* (Oxford, Oxford University Press, 1999).

24 H. F. Howson, *London's Underground* (London, Ian Allen, 1986), p. 14.

25 Jack Simmons, *The Victorian Railway* (London, Thames & Hudson, 1991), p. 24.

26 *Country Homes: The Official Guide of the Metropolitan Railway comprising Residential Districts of Buckinghamshire, Hertfordshire, and Middlesex* (London, Metropolitan Railways, 1909), p. 1.

27 J. P. Thomas, *Handling London's Underground Traffic* (London, London's Underground, 1928), p. 216.

28 *London's Latest Suburbs: An Illustrated Guide to the Residential Districts Reached by the Hampstead Tube* (London, Charing Cross, Hampstead and Highgate Railway, 1910), p. 8.

29 G. A. Selcon, 'Pullman cars on the Metropolitan Railway', *The Railway and Travel Monthly* 1: 3 (July 1910), 228.

30 T. W. H. Crosland, *The Suburbans* (London, John Long, 1905), p. 31. For a brief discussion of Crosland see John Carey, *The Intellectuals and the Masses: Pride and Prejudice Among the Literary Intelligentsia, 1880–1939* (London, Faber, 1992), pp. 57–9.

31 Crosland, *The Suburbans*, pp. 30 and 31.

32 Crosland, *The Suburbans*, p. 34.
33 Quoted in Oliver Green, *The London Underground: An Illustrated History* (London, Ian Allen, 1987), p. 31.
34 Thomas, *London's Underground Traffic*, p. 188.
35 Oliver Green, *Underground Art: London Transport Posters 1908 to the Present* (London, Studio Vista, 1990), p. 25.
36 Oliver Green, *The London Underground: An Illustrated History* (London, Ian Allen, 1987), p. 31.
37 Simmons, *Victorian Railway*, pp. 213 and 310.
38 Cited in Wolfgang Schivelbusch, *The Railway Journey: The Industrialization of Time and Space in the Nineteenth Century* (Leamington Spa and New York, Berg, 1986), p. 37.
39 Quoted in Schivelbusch, *The Railway Journey*, p. 68. For an account of the rise of railway reading that indicates the importance of women as consumers of the new fiction, see Rachel Bowlby, *Just Looking: Consumer Culture in Dreiser, Gissing and Zola* (London and New York, Methuen, 1985), pp. 86–9.
40 Georg Simmel, 'Soziologie des Raumes', quoted in Benjamin, *Charles Baudelaire*, pp. 37–8.
41 This phrase was used by Benjamin Gastineau to describe the visual experience of the *chemin de fer* in 1861; quoted in Schivelbusch, *The Railway Journey*, p. 60.
42 Day, *London's Underground*, p. 45.
43 Crosland, *The Suburbans*, p. 33.
44 For details of the growth of advertising on the tube see Thomas, *London's Underground Traffic*, p. 135; J. Graeme Bruce, *Tube Trains Under London: An Illustrated History* (London, London Transport, 1968), p. 10; and T. R. Nevett, *Advertising in Britain: A History* (London, Heinemann, 1982), pp. 57–111.
45 Elizabeth Wilson, *The Sphinx in the City: Urban Life, the Control of Disorder and Women* (London, Virago, 1991), p. 24.
46 For Christine Buci-Glucksmann a 'cult of images' is a defining characteristic of modernity. See her 'Catastrophic utopia: the feminine as allegory of the modern', *Representations* 14 (spring 1986), 221.
47 I have focused upon the male Imagists because I believe that the two female Imagists – H. D. and Amy Lowell – represent rather different tendencies within the group, especially in response to the urban experience. On H. D.'s relation to Imagist theory see Cyrena N. Pondrom, 'H. D. and the origins of Imagism', in Susan Stanford Friedman and Rachel Blau Du Plessis, eds, *Signets: Reading H. D.* (Wisconsin, University of Wisconsin Press, 1990). On Lowell and Imagism see Gillian Hanscombe and Virginia L. Smyers, *Writing for Their Lives: The Modernist Women 1910–1940* (London, Women's Press, 1987). Also see my 'Amy Lowell and H. D.: the other Imagists', *Women: A Cultural Review* 4: 1 (summer 1993), 49–59 and 'Unrelated beauty:

polyphonic prose and the Imagist city', in Adrienne Munich and Melissa Bradshaw, eds, *Presenting Amy Lowell: Critical Essays* (New Jersey, Rutgers University Press, forthcoming).

48 Richard Aldington, *The Complete Poems of Richard Aldington* (London, Allen Wingate, 1948), p. 49.

49 Richard Aldington, 'The poetry of F. S. Flint', *The Egoist* (1 May 1915), 80–1.

50 F. S. Flint, 'Tube' in *Otherworld Cadences* (London, Poetry Bookshop, 1920), p. 36.

51 Behind Flint's trope is Pound's early Imagist theory which viewed the 'spirit' animating both poet and language as akin to electricity: 'words are like great hollow cones of steel ... charged with a force like electricity': Pound, 'I gather the limbs of Osiris', *Selected Prose*, p. 34.

52 T. E. Hulme, *Further Speculations*, ed. S. Hynes (Minneapolis, University of Minnesota Press, 1955), p. 99.

53 One response was Pound's notion of the 'aristocracy' of the artist; see Ezra Pound, 'The new sculpture', *The Egoist* (February 1914), 68. For a discussion of the political affiliations of modernist 'classicism' see Alan Robinson, *Poetry, Painting and Ideas 1885–1914* (London, Macmillan, 1985), ch. 4.

54 Georg Simmel, 'The metropolis and mental life' in *The Sociology of Georg Simmel*, trans. and ed. Kurt H. Wolff (New York, Free Press, 1950), p. 418.

55 Clive Bloom, 'Introduction' to *Literature and Culture in Modern Britain*, vol. 1: *1900–1929* (London, Longman, 1993), p. 19.

56 For a consideration of the overlap between Imagism and the views of Dora Marsden, editor and founder of *The Egoist*, see my 'Dora Marsden and *The Egoist*: "our war is with words"', *English Literature in Transition 1880-1920* 36: 2 (1993), 179–96.

57 See Robert von Hallberg, 'Libertarian Imagism', *Modernism/Modernity* 2: 2 (1995), 63–79. For further consideration of the notion of modernist movements see David Peters Corbett and Andrew Thacker, 'Cultural formations in modernism: movements and magazines 1890–1920', *Prose Studies* 16: 2 (August 1993), 84–106. The most incisive recent discussion of modernist little magazines is that of Mark S. Morrisson, *The Public Face of Modernism: Little Magazines, Audiences, and Reception 1905–1920* (Madison, University of Wisconsin Press, 2001).

58 'Preface' to *Some Imagist Poets 1916: An Annual Anthology* (Boston and New York, Houghton Mifflin, 1916).

59 Hobsbawm notes how the City of London was the 'source of the world's capital, nerve-centre of its international trading and financial transactions until the 1920s'. See Eric Hobsbawm, 'Labour in the great city', *New Left Review* 166 (November–December 1987), 86.

60 Jeffrey Richards and John M. MacKenzie, *The Railway Station: A Social History* (Oxford, Oxford University Press, 1986), p. 166.

61 Richards and MacKenzie, *The Railway Station*, p. 167.
62 John Gould Fletcher, 'London Excursion' in *Some Imagist Poets: An Anthology* (Boston and New York, Houghton Mifflin, 1915), pp. 39–49. I am indebted to Bryony Randall's reading of the modernity of this poem in her hitherto unpublished paper 'John Gould Fletcher's city aesthetic: "London Excursion"'.
63 John Gould Fletcher, *Life Is My Song* (New York and Toronto, Farrar & Rinehart, 1937), p. 133.
64 John Gould Fletcher, 'London Excursion', p. 40.
65 John Gould Fletcher, 'From Babel's Night', *Coterie* 4 (Easter 1920), 37.
66 F. S. Flint, 'Accident', in *Some Imagist Poets* (1915), pp. 58–9.
67 Richard Aldington, 'The Eaten Heart', in *An Imagist Anthology 1930* (London, Chatto & Windus, 1930), pp. 3 and 5.
68 Wilson, 'The invisible *flâneur*', p. 100.
69 Simmons, *The Victorian Railway*, p. 334.
70 Buck-Morss, 'The *flâneur*, the sandwichman and the whore', p. 102. Also see Walter Benjamin, *The Arcades Project*, trans. Howard Eiland and Kevin McLaughlin (Cambridge, MA, Belknap–Harvard University Press, 1999), convolute M, p. 432.
71 Virginia Woolf, *Jacob's Room* (Harmondsworth, Penguin, 1992), p. 23.
72 Day, *London's Underground*, p. 45.
73 'Don'ts for holiday travellers', *The Railway and Travel Monthly* 1: 5 (September 1910), 395.
74 Simmons, *The Victorian Railway*, pp. 334–5.
75 Sandra M. Gilbert, 'Soldier's heart: literary men, literary women, and the Great War', in Elaine Showalter, ed., *Speaking of Gender* (London and New York, Routledge, 1989).
76 Lisa Tickner discusses the visual impact of the suffragette crowds in her *The Spectacle of Women: Imagery of the Suffrage Campaign 1907–14* (London, Chatto & Windus, 1987).
77 Quoted in Tickner, *The Spectacle of Women*, p. 46.
78 Elizabeth Wilson, *Sphinx in the City*, p. 7.
79 Wyndham Lewis, *Blasting and Bombardiering* (1937) reprinted (London, Calder & Boyars, 1967), pp. 78–9.
80 Ezra Pound, 'A few don'ts by an Imagiste' (1913) reprinted in *Imagist Poetry*, ed. Peter Jones (Harmondsworth, Penguin Books, 1972), p. 130.
81 Wilson, 'Invisible *flâneur*', p. 109.
82 Ezra Pound, 'Vorticism' (1914) reprinted in *Gaudier-Brzeska: A Memoir* (1916) (New York: New Directions, 1970), pp. 86–7. The Paris Metro had first opened in 1900 and La Concorde was, and is still, on one of the original lines of the network.
83 Pound, 'Vorticism', p. 87.
84 Pound, 'Vorticism', p. 89. The most thorough discussion of Pound's attention to typography and punctuation in the poem is by Randolph Chilton and Carol

Gibertson, 'Pound's "Metro" hokku: the evolution of an image', *Twentieth Century Literature* 36: 2 (1990), 225–36.

85 See Noel Stock, *The Life of Ezra Pound* (Harmondsworth, Penguin Books, 1974), p. 170.

86 Maud Ellmann, *The Poetics of Impersonality: T. S. Eliot and Ezra Pound* (Brighton, Harvester Press, 1987), p. 146.

87 Pound, 'Vorticism', p. 87.

88 C. F. G. Masterman, *The Condition of England* (London, Methuen, 1909), p. 121.

89 Pound, 'Vorticism', p. 92.

90 Ezra Pound, *ABC of Reading* (London, Routledge & Kegan Paul, 1934), p. 52. Arguably Vorticism's espousal of 'dynamism' only derived from the art proclaimed in a number of contemporary Futurist manifestos, and which explicitly referred to the dynamism of transport. See 'Futurist painting: technical manifesto 1910', in *Futurist Manifestos*, ed. Umbro Apollonia (London, Thames & Hudson, 1973).

91 Pound, 'Vorticism', p. 89.

92 As Brooker notes, most of the poems in *Lustra* were composed earlier, between April 1913 and March 1915; see Peter Brooker, *A Student's Guide to the Selected Poems of Ezra Pound* (London, Faber, 1979), p. 82.

93 Ezra Pound, 'Dans un Omnibus de Londres' in *Collected Shorter Poems* (London: Faber & Faber, 1968), p. 160. According to Ruthven, Pound had tried, and failed, to translate the poem into English, before telling Harriet Monroe, in whose magazine *Poetry* it was first published in 1916, that she would have to take it in the original. See K. K. Ruthven, *A Guide to Ezra Pound's 'Personae' (1926)* (Berkeley and Los Angeles, University of California Press, 1969), p. 57.

94 Ezra Pound, *Collected Shorter Poems*, p. 82.

95 Richard Aldington, 'Modern poetry and the Imagists', *The Egoist* (1 June 1914), p. 202.

96 Kenner discusses Pound's interest, during his Imagist period, in an 'aesthetic of glimpses' but without connecting this strategy with an urban environment. See Hugh Kenner, *The Pound Era* (Berkeley and Los Angeles, University of California Press, 1971), pp. 69–71.

97 T. E. Hulme, *Speculations: Essays on Humanism and the Philosophy of Art*, ed. Herbert Read (1924) (London: Routledge, 1987), pp. 134 and 136.

98 Hulme, *Images*, quoted in Jones, *Life and Opinions of T. E. Hulme*, p. 181.

99 Hulme, *Further Speculations*, p. 86.

100 Hulme, *Further Speculations*, pp. 79 and 80.

101 Hulme, *Speculations*, p. 137.

102 Jonathan Raban, *Soft City* (London, Collins Harvill, 1988), p. 248.

103 Raban, *Soft City*, pp. 248–9.

104 Wilson, 'The invisible *flâneur*', p. 110.

Ulysses, joggerfry and the Hibernian metropolis

In 'After the Race', one of the seemingly slighter stories in *Dubliners*, Joyce uses the setting of a motorcar road race to consider elliptically the modernity of Ireland. The Gordon Bennett Race, held on 2 July 1903 on a circuit to the west of Dublin, was the first major international road race to be held in Britain or Ireland.[1] In Joyce's story the cars represent a sense of European modernity to which Dublin, a colonial city of the British Empire at that time, cannot easily be assimilated. The cars come 'scudding in towards Dublin' from the surrounding countryside and, as onlookers gather to watch them, Joyce carefully indicates what these cars represent: 'through this channel of poverty and inaction the Continent sped its wealth and industry'.[2] The speeding cars embody a power and status that Joyce perceived to be lacking in Dublin; equally, their movement is in marked contrast to the 'inaction' of his native city. Famously, Joyce claimed that the aim of the stories in *Dubliners* was to highlight the debilitating 'paralysis' of Dublin.[3] In 'After the Race' that paralysis is depicted in the character of the Irish protagonist Jimmy Doyle, a well-educated but feckless young man who is the friend of Charles Ségouin, the wealthy French owner of a car which has just finished second in the race. Doyle is about to invest a large sum of money in Ségouin's motor business, and Joyce seems to suggest, as the story concludes with Doyle losing miserably at cards to Ségouin's friends, that this investment will most likely lead to 'poverty and inaction' rather than 'wealth and industry'.

Along with investing his money, and being seen by Dublin friends 'in the company of these Continentals', Doyle is excited by the car journey itself: 'Rapid motion through space elates one.'[4] The motorcar, a blue French model, epitomises European modernisation and capital to Doyle; his pleasure in the power of the technology is linked to a delight in modern life itself as the 'journey laid a magical finger on the genuine pulse of life and gallantly the machinery of human nerves strove to answer the bounding courses of the swift blue animal'.[5] Despite the characterisation of Doyle as a rather shallow enthusiast for motoring, the excitement of the car's 'rapid motion through space' is clearly associated by Joyce with a European modernity to which he, as a writer, aspired. Joyce is much more attracted by the flux of modernity than Forster or even the Imagists. Partly, Joyce associates this excitement as something lacking in the paralysed Dublin of his youth; but, while sympathetic to the fluidity of modern life, Joyce still retains a deep attachment to place. 'After the Race' indicates, in miniature, how Joyce's modernity was informed by a politics of space. How far, asks Joyce, is the urban space of Dublin capable of grasping the elation of rapid motion while remaining a colonial city? This chapter shifts attention to the period of – what is sometimes called – high modernism: 1922 saw the publication of such now-canonical texts as *Ulysses* and T. S. Eliot's *The Waste Land*.[6] My focus, however, is upon the geographical rather than temporal differences that inform the modernism of Joyce's *Ulysses*. This chapter investigates, therefore, how *Ulysses* offers a distinctive set of 'representational spaces', drawing upon Dublin's status as a colonial city within the orbit of the British Empire.[7]

In *Ulysses* we have a text that was at the heart of the exploration of the relations between space, place and modernity. The novel seems to typify Harvey's overall assessment of how modernism 'explored the dialectic of place versus space, of present versus past, in a variety of ways', and while 'celebrating universality and the collapse of spatial barriers, it also explored new meanings for space and place in ways that tacitly reinforced local identity'.[8] Joyce himself appeared well aware of this particular dialectic: 'For myself, I always write about Dublin, because if I can get to the heart of Dublin I can get to the heart of all the cities of the world. In the particular is contained the universal.'[9] In depicting Dublin in *Ulysses*, Joyce, unlike Forster, does not utilise a nostalgic sense of place as a refuge from modernity; rather, Joyce 'actualises space', in de Certeau's terms. The production of the textual spaces of the novel is always linked to movement and history, so that *Ulysses* does not so much reinforce the 'local identity' of place as unlock particular sites to the scrutiny of a spatial politics. *Ulysses* utilises both of de Certeau's spatial discourses, those of the map and the tour. Joyce's overall conception, however, appears closer to Doreen Massey's sense of place as 'numerous social relationships stretched over space'.[10] In other words, *Ulysses* (*U* hereafter) does interrogate the dialectic of place versus space, map and tour, but the effect is to show how space and what Bloom calls 'joggerfry' (geography) (*U*, p. 56[11]) are always informed by social and political relationships, in particular those between Britain and Ireland.

However, any spatialised interpretation of *Ulysses* must recognise that the text is concerned with more than just how the city is represented. Just as the city of Dublin opens out, in Joyce's view, to all other world cities, so Dublin space, as captured in Joyce's textual space, is a place where many other spaces interact. As Lefebvre argues, all forms of social space are 'traversed by myriad currents' (*PS*, p. 88) such that local, regional and national spaces are constantly assuming new forms of relationship with one another. I differ somewhat from some previous accounts of the significance of space in *Ulysses* by emphasising how inner psychic space there relies upon exterior social reality.[12] Drawing upon theories such as those of de Certeau and Lefebvre reveals more fully the complex form of Joyce's spatial vision in *Ulysses*. Criticism has focused upon Joyce's fastidious desire to portray Dublin by means of a topographical verisimilitude, or upon his employment of an 'epic geography' drawn from the Mediterranean world of Homer, but has mainly ignored conflicts over the political meanings of space and place.[13] However, the spatial politics of Joyce's text is found not only in the portrayal of space and place – Dublin, Ireland, the world – but in the textual space of his novel, and in the interrelationships between material and metaphorical senses of space. As in previous chapters I am concerned both with how Joyce represents the social spaces of Dublin and, equally, with how the spatial form of Joyce's text is shaped by the social spaces of the city and the nation.

This chapter examines first how a specific site in the city, a public statue of Lord Nelson, encodes various views of national and colonial space in *Ulysses*. The discussion turns then to the nature of some of the movements through the city, particularly those courtesy of the Dublin United Tramways Company, and demonstrates how this urban transportation system informs Joyce's stylistic use of the rhetorical feature of chiasmus. The peregrinations of Bloom and others constitute another manifestation of what I have been tracing as a movement through modernity, a journey that commences in a city and eventually concludes in the 'representational space' of the novel. To that end I highlight the social and political nature of the modernity Joyce surveys in his text, showing the conflictual nature of its spaces, particularly in relation to colonial power in the 'Wandering Rocks' episode and, picking up the arguments I traced in chapter 3, the spatial politics of the gaze in Bloom's voyeurism throughout the book.

The Hibernian Nelson

The central sequence of episodes from 'Lotus Eaters' to 'Wandering Rocks' is where Joyce concentrates his attention on the spaces of the modern city. Clearly, many other *topoi* have a significant presence in the novel, but in order to limit analysis to manageable proportions I have chosen to focus upon these episodes. The opening words of the seventh episode, 'Aeolus', situate the novel not just within Dublin but 'IN THE HEART OF THE HIBERNIAN METROPOLIS' (*U*, p. 112). The previous episode, 'Hades', has seen Paddy Dignam buried, killed by a heart attack, and the episode is replete with puns and references to the organ of the heart. 'Hades' concerns a journey by carriage across the city to Dignam's funeral in Glasnevin Cemetery. Dublin is viewed from the inside of the carriage, another kind of 'placeless place' in Foucault's terms, a form of negative heterotopia leading to the cemetery.[14] We shift then, in 'Aeolus', from the defunct organ of Dignam's corpse into the throbbing heart of the city.

Getting to the heart of this city was one of Joyce's stated aims, from the earliest stories in *Dubliners* to the personification of the river Liffey as Anna Livia Plurabelle in *Finnegans Wake*. In *Ulysses* Joyce famously aimed 'to give a picture of Dublin so complete that if the city one day suddenly disappeared from the earth it could be reconstructed out of my book'.[15] But Joyce was always ambivalent about Dublin's status as a metropolis, as the stress on 'paralysis' in *Dubliners* indicates. One typically equivocal description occurs in 'After the Race', when Joyce describes Dublin as wearing 'the mask of a capital'.[16]

To situate 'Aeolus' in the *heart* of this city masquerading as a capital is, therefore, a typically mordant comment by Joyce. All of the headlines in this episode offer a parodic commentary on the rhetorical wind of Dublin newspapers and journalists. Is the claim evinced for Dublin's metropolitan status thus merely another example of journalistic hot air? Joyce situates Leopold Bloom at the city's heart, but he seems equivocal over the status of this city: Hibernian heart, but not the central organ of a nation, since London is still the imperial capital of Ireland. It is interesting to note that the headlines for 'Aeolus' were added only late in the composition of the text. The episode was first published in *The Little Review*, in October 1918, and Michael Groden suggests that the headlines were added as late as August 1921.[17] It is also significant that the first headline added omitted the word 'Hibernian'. Subsequent revisions to produce the 'heart of the *Hibernian* metropolis' allowed Joyce an additional resonance to the theme of national space that is the main topic of political discussion in the episode. In a letter to his brother, Joyce described Dublin as being both 'a capital for thousands of years' and 'the "second" city of the British Empire'.[18] The geopolitics of this dual status for the city is a topic windily discussed throughout 'Aeolus'. At one point Professor MacHugh provides a dreamy description of Ireland from Dan Dawson's speech, headlined 'ERIN, GREEN GEM OF THE SILVER SEA' (p. 119), and concluding with 'Our lovely land', to which Bloom innocuously replies 'Whose land?' (p. 120). The professor describes this as a '[m]ost pertinent question ... [w]ith an accent on the whose' (p. 120). This episode is the most detailed exploration of national identity so far, with the politics of home rule obsessing newspaper debate in this period in Ireland.[19] Bloom, perhaps inadvertently again, contributes to this by wanting to place an advertisement in *Freemans Journal* for Alexander Keyes, a merchant, that utilises a visual and verbal pun on the parliament of the Isle of Man, the House of Keys, and which, as Bloom notes, has the '[i]nnuendo of home rule' (p. 116). The innuendo relies upon the fact that the Isle of Man, unlike the island of Ireland at this time, had a parliament with some independence from London governance.

Aside from the characters' discussion of national identity, it is in the references to specific sites within the city that Joyce most evocatively explores Dublin's equivocal metropolitanism. The opening of 'Aeolus' is worth dwelling upon for how Joyce portrays the 'Hibernian metropolis', and how the material spaces of the city interact with a variety of metaphorical meanings for particular locations. The opening paragraph strikes a seemingly prosaic note in its vision of city trams departing from Sackville (now O'Connell) Street, in the centre of Dublin. This paragraph was another late addition in Joyce's compositional process[20] – which should warn us that this piece of apparently inconsequential detail masks a more significant set of connotations:

> Before Nelson's pillar trams slowed, shunted, changed trolley started for Blackrock, Kingstown and Dalkey, Clonskea, Rathgar and Terenure, Palmerston park and upper Rathmines, Sandymount Green, Rathmines, Ringsend, and Sandymount Tower, Harold's Cross. The hoarse Dublin United Tramway Company's timekeeper bawled them off:
>
> — Rathgar and Terenure!
> — Come on, Sandymount Green!
>
> Right and left parallel clanging ringing a doubledecker and a singledeck moved from their railheads, swerved to the down line, glided parallel.
>
> — Start Palmerston Park! (p. 112)

There is much that is significant about this list of destinations in Dublin, other than simply to add urban verisimilitude.[21] In the initial reference to Nelson's pillar we see how Dublin's geographical heart, and the centre of its transportation system, concretely connects the city to its imperial ruler London. A contemporary advert for the Dublin United Tramways Company has a picture of Nelson's pillar with the caption: 'The Nelson Pillar. The Centre of Dublin Tramway System'.[22] It is worth exploring, then, the cultural history of Nelson's pillar to uncover the wider significance of its place in Joyce's textual space.[23]

Nelson's pillar was a statue of Lord Nelson placed atop a large Doric column in the centre of Sackville Street, and was the first monument to be erected in this central site in Dublin (see figure 5). It stood an imposing height of 134 feet and had been erected in 1808–9 by public subscription. Nelson, though mortally wounded, had been the English victor over the French and Spanish fleets at the battle of Trafalgar in October 1805. The pillar's architect was an Englishman, William Wilkins, a leading protagonist of the Greek revival style in architecture. Wilkins built a number of public works in this same neo-classical style, including the National Gallery in London and another, later, column commemorating Nelson in Great Yarmouth, Norfolk.[24] Although Wilkins is the named architect there is some dispute over the extent to which he had overseen the building of the column. Some commentators suggest that the final design owes more to the Armagh-born architect Francis Johnston, who was in charge of the actual construction of the pillar. Johnston was to be the architect responsible for building the adjacent General Post Office building in Sackville Street, in 1815.

One 1907 guide to Dublin drew attention to the fact that, although the city is 'well supplied with memorials of those Irishmen whom their country has delighted to honour', the most striking monument, 'here as in London', is to 'the great English admiral whose name is a passport to the enthusiastic admiration of all English-speaking peoples'.[25] The pillar was decorated with the names and dates of Nelson's victories, and could be climbed from the inside to offer 'a magnificent panorama of Dublin and its surroundings'.[26] The statue inserts Dublin deep within the colonial space of the British capital, the heart of which is often seen as Trafalgar Square with its Nelson's Column. The metropolitan centre of Ireland is thus, paradoxically, not an Irish centre; or, to adopt a phrase from Derrida, the centre is not the centre: the Irish metropolis has its governmental centre in London.[27]

Interestingly, the Nelson statue in London was built only some thirty years after the completion of the Dublin Nelson, emphasising its importance as an image of Irish union and loyalty with Britain. In a discussion of other Dublin landmarks in Joyce's 'The Dead', Luke Gibbons argues that '[p]ublic monuments are expressions of official memory, and bear witness to the power of the state to legitimise its triumphant version of the past ... [b]y their imposing presence, and their control of public space'.[28] Built soon after the 1801 Act of Union, Nelson's pillar symbolised both the political union of Britain and Ireland and the dominance

5 The Nelson pillar, Sackville Street, Dublin

of the British State's view of history and its 'official memory'. Writing in 1909, one Irish commentator noted that 'Nelson's pillar standing in the middle of O'Connell street represents the dominion of England; O'Connell's statue at the end typifies the Irish people.'[29] Pointedly, 'Aeolus' contains no reference to the statue of the nineteenth-century nationalist leader Daniel O'Connell, unlike the 'Hades' episode in which the funeral cortège passes beneath 'the hugecloaked Liberator's form' (p. 90), linking O'Connell to the theme of death in the episode. In the newspaper office, in 'Aeolus', one character, Professor MacHugh, describes the Irish as 'liege subjects of the catholic chivalry of Europe that foundered at Trafalgar' (p. 128) due to Nelson's victory. Joyce's employment of Nelson's pillar thus demonstrates how the literary form of his text is implicated in spatial relations of power–knowledge, as the column shadows the opening and closing of the episode, forming an impression of the city as being under surveillance.[30] Commenting on the impression of 'phallocratic' authority conveyed by skyscrapers and public buildings, Lefebvre notes: 'Verticality and great height have ever been the spatial expression of potentially violent power' (*PS*, p. 98). The Nelson pillar was thus, in

Lefebvre's terms, a clear manifestation of an official 'representation of space': a monument conceived to signify British imperial rule in Ireland.

The status of Nelson's pillar as a symbol of British power was still recognised in March 1966 when, as part of unofficial commemorations of the Easter Rising of 1916, a dissident faction within the IRA blew up part of the pillar.[31] The *Irish Times* reported the explosion with a headline that could slip unobtrusively into 'Aeolus': 'NELSON DEFEATED IN THE LONG RUN. ABUSED FOR 157 YEARS'.[32] Throughout its existence the pillar was the site of discord between different versions of Anglo-Irish history. Soon after the news of Nelson's demise at Trafalgar had reached Dublin, the alderman of the city argued for 'the expediency of some speedy and practicable measure to compliment the memory of Lord Nelson'[33] According to Henchy great grief was displayed in Dublin at the loss of Nelson, with the *Dublin Evening Post* noting how 'all breathe a noble unanimity of affection for the memory of the departed hero'[34] Some of the first dissenting voices occurred just after the completion of the statue in 1809, such as the following attack in the *Irish Magazine*:

> The statue of Lord Nelson has been placed on the column in Sackville Street dedicated to his memory. We never remember an exhibition that has excited less notice, or was marked with more indifference on the part of the Irish public, or at least that part who pay the taxes and enjoy none of the plunder ... English dominion and trade may be extended, and English glory perpetuated, but an Irish mind has no substantial reasons for thinking from the history of our connexion that our prosperity or our independence will be more attended to, by our masters than if we were actually impeding the victories, which our valor as personally effected ... We have changed our gentry for soldiers, and our independence has been wrested from us, not by the arms of France, but by the gold of England. The statue of Nelson records the glory of a mistress and the transformation of our Senate into a discount office.[35]

Towards the end of the nineteenth century attempts were made by critics to have the pillar removed from its original site. Objections to the column's appearance, or that it hindered traffic, soon shaded into political disagreements over the meanings attached to the monument. In 1876 a Dublin Corporation meeting proposed that the pillar should go, and in 1891 the Irish Parliamentary Party at Westminster tried unsuccessfully to introduce a bill to topple Nelson. The attorney-general of Ireland and the Unionist members for Ulster, keen to maintain the links between Britain and Ireland at a time of intense agitation over home rule, opposed them.

Others defended the political connotations of the statue, notably W. B. Yeats in a 1923 speech while a senator in the *Dáil* (the Irish Parliament). Yeats characteristically retained an aesthetic reticence regarding the Doric column: 'It represents the feeling of Protestant Ireland for a man who helped to break the power of Napoleon. The life and work of the people who erected it are a part of our tradition. I think we should accept the whole part of this nation and not pick and choose. However, it is not a beautiful object.'[36] Yeats alludes to the fact that those who subscribed in order for the statue to be built, led by the alderman of Dublin, were mainly Protestant merchants delighted that Nelson's victory at Trafalgar entailed the end of the trade blockade between the British Isles and the Continent. Writer and friend of Joyce Padraig Colum showed a more inventive approach in his suggestion: 'If there are people who do not like the idea of a British admiral dominating a Dublin street, give him another arm and call him by another name – Robert Emmet perhaps.'[37]

Nelson's pillar is thus a good symbol of how Joyce interrogates the spatial history of his home city, drawing upon the complex relations between Ireland and England captured in the monument. In particular the notion that the 'heart', or centre, of Dublin and Ireland is markedly English perhaps appealed to Joyce as an example of the spatial ironies of Irish history. Joyce, therefore, is discretely referring to a spatial politics, a struggle over the meanings and memories encoded in particular places and locations. The occasion that the intended destruction of Nelson's pillar was to commemorate, the Easter Rising of 1916, has as its most potent monument the nearby General Post Office, occupied by the rebels led by Patrick Pearse, and destroyed in the ensuing fighting. The GPO building, designed by Johnston, the architect who had overseen the erection of Nelson's pillar, was another site redolent of the Anglo-Irish ascendancy, utilising the same neo-classical style (figure 6).

It is not surprising, then, to find that the second paragraph of 'Aeolus' refers to the General Post Office and 'His Majesty's vermilion mailcars, bearing on their sides the royal initials, E.R.' (p. 112). The headline here is a further reminder of the geopolitical links between Dublin and London: 'THE WEARER OF THE CROWN' (p. 112).[38] Joyce shows how material spaces, such as a building or a monument, are always the site of political disagreements: his text functions as a representational space, staging the contradictory meanings of the pillar's connection to London. Composed after the 1916 Easter Rising the headlines and references to significant sites of spatial history such as Nelson's pillar and the General Post Office demonstrate how 'Aeolus' maps out the potential conflict embedded in the very streets of the city. We should, however, be alert to any simplistic reading of Joyce's

GPO building and the Nelson pillar, Easter 1916

references to the colonial architecture of Dublin, for this episode is dominated by the windy rhetoric of nationalism, and by the theme of frustration. In *The Odyssey* Odysseus's journey home is thwarted when, nearing Ithaca, his crew open the bag of winds given him by Aeolus to assist his journey. The ship is blown back to the land of Aeolus, who refuses to help Odysseus a second time. This might be taken as Joyce's acerbic comment on the 1916 revolutionaries who, approaching home rule, get parts of the city blown up and thus undermine their lofty ambitions. The sense offered in Linati's schemata for this episode only emphasises this ambivalence towards the 1916 rising: 'The Derision of Victory'.[39]

When, at the end of the episode, the pillar forms the subject of Stephen Dedalus's 'vision' of the two old Dublin women, Anne Kearns and Florence MacCabe, his saucy parable serves only to emphasise that derision. The two women 'want to see the views of Dublin from the top of Nelson's pillar' (p. 139) and so they scale the pillar to obtain the view, only to get a kind of vertigo at the summit. Their view is described as a 'Pisgah' one, an unachievable vision, and is linked to the theme of a disappointed Moses in the promised land.[40] Though the women – symbolising an unheroic 'mother Ireland' – scale the phallic pillar of British imperial power their achievement is derisory. They fear the pillar might collapse, an allegory of the fall of the

British Empire, linked in the episode to the fall of another empire, that of the Romans. But the fall does not occur, the pillar remains, and the women gobble plums, throwing the stones over the railings to the streets below.

This is an example of de Certeau's 'ways of operating', where the Dublin women appropriate a public monument for their own idle devices. But Joyce reminds us that Nelson, the 'onehandled adulterer'[41] (p. 142), still retains his lofty presence in the city, compromising any symbolic victory attached to the women's conquest of the pillar. The vertical imperial power embodied in Nelson is in stark contrast to the horizontal lack of electrical power in the Dublin trams, 'still, becalmed in short circuit' (*U*, p. 143), at the end of 'Aeolus'. The static image of imperial power, the panopticon-like gaze of the coloniser, finds a mirror image in this paralysis of the Dublin trams.

Later, in the 'Wandering Rocks' episode, we encounter the two women again, walking 'through Irishtown along London bridge road' (p. 232). Michael Seidel notes that this incident shows how Joyce uses place-names to 'expand the world of the narrative'.[42] Such instances not only expand narrative space, but show the interconnectedness of material and metaphorical space, that of the city and that of the text. The place-names link Irish space to the English capital, in a similar way similar to the presence of Nelson in both Dublin and London. Toponyms also grant Joyce another technique by which to transform material into textual space: the solid stone of Nelson's pillar is converted into the song *The Death of Nelson*, sung by the one-legged sailor in 'Wandering Rocks' (*U*, p. 216). This English nationalist song repeats the phrase 'For England home and beauty', mirroring the Irish nationalist songs that circulate in the novel. This textual Nelson is then connected back to a material space when the sailor 'growled at the area of 14 Nelson street' (*U*, p. 239), a street bisecting Bloom's home in Eccles Street.

Joyce here illustrates the historical intertwining of Ireland and England as it is mapped in the streets of Dublin. Place-names tell historical stories about specific loci, stories revealing acts of possession and dispossession by those with the power to grant new names to old locations.[43] If Joyce's reference to Nelson expands his narrative into the space of colonial relations of power, it also seems to suggest an acceptance of the diverse nature of the place that one inhabits, rather than a rejection of elements of the relationship with England that one dislikes. Joyce, in a conversation with Budgen about Irish home rule, suggested that 'Ireland is what she is ... and therefore I am what I am because of the relations that have existed between England and Ireland. Tell me why you think I ought to wish to change the conditions that gave Ireland and me a shape and a destiny?'[44] In other words, Joyce's use of Nelson demonstrates Lefebvre's view that 'social spaces interpenetrate one another and/or superimpose themselves on one another' (*PS*, p. 86). The 'shape' of Joyce the Irishman was formed in the spatial relationships produced *between* England and Ireland, London and Dublin. Similarly, Joyce's text takes shape, not just as a set of representational spaces but as a profound attempt to compose an innovative textual space as a prelude to a new national space.[45]

Joyce's desire to carve out a new representational space from the rough-hewn material of Irish history is also confirmed by the appearance of 'Aeolus' on the page. The typographical layout of the newspaper headlines is one of the first great shocks of *Ulysses*: the headlines disrupt the flow of narrative and foreground the space of the printed page itself. Are we now reading a novel or a newspaper? Is there a voice or character to be attached to these headlines?[46] This textual space queries our reading practices, as we shuffle from headline to the story below, sometimes bewildered as to their putative connection. The morphology of 'Aeolus' is thus, literally, an 'in-between space' – between newspaper and novel, or – to utilise a suitably Joycean pun – between Nelson's British column and the printed columns of the Irish press. In this way Joyce's text is engaged in a spatial politics, trying to produce a representational space that contests official and monolithic versions of national identity, whether British or Irish.

Reconstructing Dublin space was, we might say, something Joyce regarded as his own prerogative, rather than that of Pearse and the other rebels of 1916: the space of his text would rebuild Dublin as something more than the 'second' city of the British Empire. Not content merely to represent the topography of Dublin, Joyce wished to transform it into the textual space of his novel. In this transformation we witness an open-ended debate over the semiotics of city sites such as the Nelson pillar, with its tangled cultural history of Anglo-Irish relations. Equally, the typographical appearance of 'Aeolus', with its questioning dialogue between two different forms of discourse, is shaped by the parallelism between the two Nelsons: the Irish Nelson and the English Nelson, or the Nelson of imperial power versus the admiral whose column afforded a fine view of Dublin for old women and was a source of many puns and parables. Joyce moves, then, into the crevices and crannies of the city in order to re-order social space into a newly constructed textual space.[47] As one of the 'Aeolus' headlines trumpets: 'ITHACANS VOW PEN IS CHAMP' (p. 142). Joyce's pen, we might add, has the final say over Nelson's phallic column.

The velocity of modern life

The static and stately presence of Nelson is in contrast to the many instances of the elation produced by motion in *Ulysses*. Travel is a significant feature of the text, with the expansive peregrinations of Odysseus in the Mediterranean being transformed into the much more circumscribed loops and circuits of citizens in Dublin. Although most of the journeys represented in the novel are by pedestrians, the Dublin trams feature in a number of significant ways. Commencing in 1865 the Dublin tram system grew until it was considered in 1904 one of the most impressive in the world.[48] Budgen noted the strong influence of the transport system on Bloom's acute sense of modern time and space: 'The discoveries of the astronomer and the mathematician have less immediate effect on this sense than the electrification of the suburban lines.'[49] A 1907 guide to Dublin praised the tram system 'which by successive extensions and improvements now renders Dublin in respect to internal communications second to no city in Europe'.[50] The same guide reprints a list of the various tramlines as part of its tour itinerary of what it terms 'the Irish metropolis'.[51] The reference to trams radiating outwards from Nelson's pillar at the start of 'Aeolus' shows Joyce exploring the vexed perception of Dublin as a modern European city. Are the Dublin trams a true indication of the modernity of Ireland, demonstrating what Bloom terms 'the velocity of modern life' (p. 672) – and what Joyce, in 'After the Race', had thought was absent from his native city?

The relationship between Ireland, modernity and modernisation has been a question of much debate from the 1990s onwards among Irish critics such as Luke Gibbons and Seamus Deane.[52] Gibbons, for example, argues that due to the impact of centuries of colonial rule, Irish history is characterised by the sorts of fragmentation and disruption normally associated with an experience of modernity caused by accelerated industrialisation. In a sense, Gibbons suggests, the 'shock of modernity' was experienced in Ireland well before the twentieth century, and this has important consequences for understanding the relation of Irish writers to modernism and modernity:

> this calls for a reversal of the standard view which presents the modernist movement – particularly as represented by Joyce and Beckett – as turning its back on the torpor of tradition in Ireland in order to embrace the exhilaration of the metropolitan avant-garde: if anything, these writers' vital contacts with mainland European culture proved productive precisely because they were carrying with them the nightmare of Irish history.[53]

One problem with Gibbons's argument, however, is the one-sidedness of his view of modernity and modernisation. The excitement and thrill of the modern, as typified by life in the metropolis, is missing from a modernity characterised only by fracture and upheaval. Seamus Deane offers a slightly different account of Irish modernity when he writes:

> To be colonial is to be modern. It is possible to be modern without being colonial; but not to be colonial without being modern. Ireland exemplifies this latter condition and presents it in such a manner that the 'traditional' and the 'modern' elements seem to be in conflict with one another, like two competing chronologies.[54]

Or competing topographies. In the representation of technologies such as the Dublin trams we gain an insight into Joyce's complex vision of the spaces of colonial modernity. If the trams in *Ulysses* represent an image of the 'velocity of modern life', they co-exist in urban space with other images of another, more 'traditional', Ireland.

Frank Budgen describes Bloom as 'a spacehound, an advertisement canvasser' or a 'spacetimehound, seeing that he sells quantities of space for periods of time'.[55] Given that trait, it is not surprising that Bloom's imagination often turns to matters of geography and transport in his native Dublin. In 'Ithaca', the episode of Bloom's homecoming, we learn more about Bloom's interest in schemes for improving Dublin's transport infrastructure. He imagines, for example, developing 'tourist traffic in and around Dublin by means of petrolpropelled riverboats ... charabancs, narrow gauge local railways, and pleasure steamers for coastwise navigation' (p. 671). Among the many desired features of Bloom's dream dwelling – 'Bloom Cottage. Saint Leopold's. Flowerville' (p. 667) – is that it be 'not less than 1 statute mile from the periphery of the metropolis, within a time limit of not more than 5 minutes from tram or train line' (p. 665). This indicates both Bloom's constant eye for new business ventures (that will probably never be ventured) and his steadfastly lower-middle class outlook, desiring to live in suburban Dublin, but with ease of access to the centre of the city. Unlike the tube in London, the growth of Dublin's trams seemed to follow, not institute, the growth of the suburbs and, from the relatively high price of tickets, the system seemed geared around white-collar workers like Bloom.[56] In Bloom's budgetary inventory for the day we note an expenditure for two tram journeys (p. 664);[57] doubtless Bloom would be distressed to learn that the Dublin trams have also that day conveyed a basket of 'fat pears and blushing peaches' (p. 219) to his wife, Molly Bloom, sent by her lover Blazes Boylan.

In 'Ithaca', Bloom also ponders a scheme to improve the transport of cattle across the city by connecting 'by tramline the Cattle Market (North Circular road and Prussia street) with the quays (Sheriff street, lower and

East Wall), parallel with the Link line railway laid (in conjunction with the Great Southern and Western railway line) between the cattle park, Liffey junction, and terminus of Midland Great Western railway' (p. 671). This idea first occurs to Bloom when the funeral cortège across the city in 'Hades' is halted by a herd of cattle (and sheep) being driven to the docks, probably for export to Liverpool to become '[r]oast beef for old England. They buy up all the juicy ones' (p. 94).

Cattle in the novel are symbolically associated with Ireland itself, starting with the description of the old woman in the first episode who brings the milk to the Martello tower as the 'silk of the kine', a feminine image for Ireland common since the eighteenth century.[58] In the second episode, would-be Irish artist Stephen is the 'bullockbefriending bard' who must take Deasy's letter about foot-and-mouth disease to the newspaper office. If the cattle represent Ireland then Bloom's thought of their export to England perhaps echoes the familiar story, much contested, of the exporting of food during the Irish famine of the 1840s. But Bloom's main consideration about the cattle is how to improve their transit to the dock-side: 'I can't make out why the corporation doesn't run a tramline from the parkgate to the quays ... All those animals could be taken in trucks down to the boats' (p. 94). This idea – which becomes an election pledge in Bloom's fantasy of power in 'Circe' (p. 452) – is better received by Bloom's fellow mourners than his other notion: municipal funeral trams as in Milan, with hearse and carriage for the corpse and mourners. Bloom thus sees the trams as a modernising technology, one that might improve Dublin's urban fabric so as to bear comparison with those of other European cities. Such a modernising impulse in Bloom is also noted when, looking out of the window of the funeral carriage in 'Hades', Bloom notices a 'pointsman' changing the tram track manually: 'Couldn't they invent something automatic so that the wheel itself much handier?' (pp. 88–9)

This Joycean geography of cattle and trams, rural Ireland and the Dublin metropolis, indicates how *Ulysses* projects a polytopic vision of Lefebvrean social space. Cattle impeding the traffic in the city might be read as rural Ireland thwarting the progress of a modernity forlornly wished for by Joyce. Another view of the co-existence of cattle and trams in the streets might query this easy association of mechanisation with modernity. Following this interpretation the cattle clogging the Dublin streets would not represent a rural refusal of the modernity of the trams; rather they might indicate a society where quite different conceptions of space and time could co-exist, in the manner of Bakhtin's description of the overlapping existence of distinct chronotopes.[59] This hybrid chronotope, tradition and modernity, could be read as a metaphor for Ireland's uneven economic development as a peripheral place on the margins of the British Empire. Marjorie Howes, for example, suggests that 'because Ireland's modernization was inextricably bound up with the uneven development of colonialism, highly modernized communications and transportation systems ... occurred in conjunction with archaic or residual social, economic, and cultural formations'[60] Dublin, for example, played a key role in the Irish economy as a port, not for its trade with the rest of the world, but due mainly to the cross-channel shipping to Britain. Bloom's thoughts about the destination of the cattle are borne out by the fact that Britain 'was the supplier of most imports and the destination of the principal exports' to and from Ireland, particularly of cattle.[61]

The different meanings attached to what occupies the streets of Dublin are replicated when critics try to link these material spaces to the expanse of the text. Jameson, for example, describes Joyce's Dublin as an 'underdeveloped village',[62] and, for the initiated, the bustling enclosed space that is the pages of this book has something of the quality of a busy village on market-day, where everyone is familiar with everyone else. For the novice reader, however, the amplitude of the text is profoundly alienating at first, and this form of alienation is most readily associated with the literature of the modern urban experience. Metropolis or village, trams or cattle, once again Joyce's text demonstrates Lefebvre's notion of the 'ambiguous continuity' between different social spaces within the same geography.

In this respect Joyce's text complicates certain of the arguments current in Ireland at the time (and since) about the relative values of the city and the country as spatial signifiers of Irish national identity. Gerry Smyth has argued that, from the eighteenth century onwards, 'modern Irish nationalism was concerned as much with issues of space (ownership of the land, landscape and identity, geography as destiny, exile, and so on) as with the temporal status of the nation'.[63] One key element, suggests Smyth, was 'the elevation of the country over the city as a signifier of Irishness', a discourse that, paradoxically, derived from English colonial views of the Irish as a backward race of disempowered rural peasants when compared to the supposedly progressive urban values of England. The authentic heart of the Irish nation was thus located in the countryside, symbolised in a figure like that of Yeats's fisherman, from whose inspiration the poet would 'write for my own race'.[64] As Smyth argues, Joyce writes against this nostalgic rural myth for the way in which it 'denied the validity of urban experience, as if city-dwelling was somehow less "Irish" than living in a village ... It was this implicit spatial politics that partly inspired Joyce's championing of the city and his exploration of different modes of Irishness.'[65] This perhaps explains why Joyce situates the heart of Hibernia within the metropolis rather than the boglands of Galway. Smyth's argument, though convincing, slightly misses the way in which Joyce's geographical vision encompasses city and country, trams and cattle, without privileging either. Joyce's 'championing of the city' is, as 'Aeolus' demonstrates, something of a qualified passion. The key point is to understand the movements between overlapping conceptions of space and place in *Ulysses*.

Joyce's cartography of urban space always gestures towards the interconnections between places and what de Certeau calls their 'spatial stories'. The tram destinations at the start of 'Aeolus' show how the spatial history of the city is embedded in many of the very names of its streets and areas. Places like Kingstown (now Dun Laoghaire) or Palmerston Park, like Nelson's pillar, indicate the political topography of Dublin. Lord Palmerston, British prime minister in the 1850s and 1860s, was known in Ireland as an absentee landlord, notorious as one who did not assist any of his tenants in the passage to America during the time of the famine.[66] The fact that one tram is headed to Sandymount Green links the material space of the city to the metaphorical space of Joyce's novel, since that tram is destined for the Martello tower encountered at the very start of the book. Dotted around the coast of Ireland (and the south-east of England), Martello towers are, like Nelson's pillar, another location telling spatial stories with strong political associations. They were constructed during the Napoleonic Wars as lookout posts for signs of possible French invasion; but for nationalist Ireland such an invasion would help its own rebellion against British rule.[67] In the first episode of *Ulysses*, 'Telemachus', Joyce emphasises the military role played by the Martello tower: the building contains a 'gunrest' and is jokingly called a 'barracks' by Buck Mulligan, while Stephen caustically comments that the rent on the tower is paid to 'secretary of state for war' (p. 17).

In this way *Ulysses* displays de Certeau's 'actualization of space', whereby the historical and political meanings of social space are always made visible. Often what appears in *Ulysses* as a version of de Certeau's mapping discourse, the static ordering in a tableau of places, is actually informed by a tour discourse, which emphasises the *movements* and histories by which a particular space or place came into being. The convoluted political history of buildings such as Nelson's pillar is one example. The list of tram destinations is another. They are instances of Joyce's 'fanatic naturalism'[68] in terms of mapping city space; but they also employ something akin to de Certeau's discourse of the tour, when material spaces produce deeply metaphorical meanings. The two trams moving in parallel at the start of 'Aeolus', 'a doubledecker and a singledeck' (p. 112) symbolise many of the other parallels in this episode and in the novel overall. In 'Aeolus' Bloom and

Dedalus travel a kind of parallel journey, with both entering the same newspaper office but without meeting. Likewise, Britain exists to Ireland as a doubledecker to a singledecker in terms of relations of power, even as they exist, geographically, alongside one another. And London and Dublin, of course, contain parallel monuments to Nelson.[69]

In another sense the motions of the Dublin trams represent an image of how the reader encounters the space of *Ulysses*, circulating through the pages backwards and forwards, coming across familiar phrases, words or characters whose significance is not developed in any linear fashion. In 'Aeolus' the narrative loops back to finish, like a tram on a return journey or Odysseus's homeward travels, with a final flourish underneath Nelson's pillar once more. But though Joyce's text exhibits this circular motion, the trams are paralysed through the electricity failure. Dublin's symbolic paralysis is, therefore, at the expense of Joyce's textual travels.

Reading *Ulysses*, as students always remind lecturers, is something of an epic journey in itself. But in many ways the analogy with a journey is an apt one for the very fact of reading the text, even though it is perhaps not a movement of 'rapid motion' that initially elates one. To travel through *Ulysses* is more akin to wandering as a stranger through a new city, as Hugh Kenner once noted:[70] at first all is a mystery, getting from A to B is slow progress, and we often enlist the help of guides or maps like the admirable Blamires to assist us. During subsequent visits the reader begins to build a cognitive map of the text: familiar themes, characters or phrases become the landmarks of a city by which we navigate our perambulations. Tried and tested routes are hereafter followed habitually; for example, we start to notice just how often the Stephen–Hamlet conjunction is raised. And just as when we walk along familiar city streets features that were once fresh become overlooked, and towers, buildings, lights and alleys are not now noticed, but ignored in favour of minor details – a church steeple, the brickwork on a particular house front, a vibrantly coloured door – thus we start to notice new textual sights, such as the betrayal theme embedded in the 'croppy boy' motif, or the blind stripling's journey across the city. There are main thoroughfares to the novel, and these are plied by the early reader, and re-plied by the re-reader, much as we might take the same bus to work each day and become, to an extent, habituated to the view from the carriage window.

Though these are spatial metaphors for the experience of reading *Ulysses*, it is also true that the text problematises the opposition between material and metaphorical space. The text is material space through which to pass, and to map these pages is to read that space, as much as reading it is to map out a path through and across it. If *Ulysses* initially appears as a bizarre form of map, the experience of reading it resembles de Certeau's sense of a plurality of tours and spatial stories. The reader's experience of the spatial form of the novel, as Frank termed it, is thus intimately connected to the kinds of movements through the social spaces of the city itself. Hugh Kenner suggests that the 'episodes' (rather than the 'chapters' of the conventional novel) structuring *Ulysses* might be traced to the impact of a city 'shaped by rapid transit' and which 'delivers its experience in discrete packets', like passengers on a tram.[71] Thus the overall narrative structure of the text owes something to moving across the social space of the city: material and metaphorical spaces interweave, showing once again Lefebvre's hypercomplex understanding of space.

Criss-cross

Joyce's choice of rhetorical tropes, particularly in 'Aeolus', often reveals interesting connections between the social space of the city and the spatial styles of *Ulysses*.[72] One interesting trope is that of the rhetorical figure of chiasmus, a form of linguistic parallelism to match the motion of the trams. Chiasmus, derived from the Greek letter *X* (*chi*), refers either to a sentence whose word order is reversed in the next sentence, or where oppositions in a sentence exchange places in a criss-cross or X shape.[73] Its effect on the reader's eye is to draw attention to the spatial layout of words on the page, imparting a distinctive spatial pattern to the typography. To these senses we can add a third usage, where chiasmus refers to a crossing over between material and metaphorical spaces.

The criss-crossing of the trams at the start of 'Aeolus' seems to produce a chiasmic description of the brewery workmen: 'Grossbooted draymen rolled barrels dullthudding out of Prince's stores and bumped them up on the brewery float. On the brewery float bumped dullthudding barrels rolled by grossbooted draymen out of Prince's store' (p. 112). In a number of other instances in the novel Joyce's use of linguistic chiasmus is prompted by the material spaces of transport. In 'Ithaca', for instance, Bloom gazes at the movements of the night stars in interstellar space while he considers the idea of wandering 'the extreme boundary of space' (p. 680) to avoid his domestic woes. The possibility of a return journey he considers less favourably because of an 'unsatisfactory equation between an exodus and return in time through reversible space and an exodus and return in space through irreversible time' (p. 680). Only a few moments after this syntactic cross-over Bloom remembers the face of his father-in-law Major Tweedy, and another image of modern transport occurs to him:

> Retreating, at the terminus of the Great Northern Railway, Amiens street with constant uniform acceleration, along parallel lines meeting at infinity, if produced: along parallel lines, reproduced from infinity, with constant uniform retardation, at the terminus of the Great Northern Railway Amiens street, returning. (*U*, p. 682)

Bloom's travels are now at an end, for he has arrived home: 'He rests. He has travelled' (p. 689). Joyce's text, however, now travels on to the space of infinity, where parallel lines meet, represented by Molly Bloom in the next, and final, episode. The movements of trams, trains and Bloom run parallel and then cross-over into the textual motion of the novel, for Joyce is perhaps aware that the reader of *Ulysses* is, by reaching these pages, a somewhat weary traveller.

Joyce seems keenly aware of the inherently spatial characteristics of the figure of chiasmus. The parallel paths of Bloom and Stephen through the city enact the same kind of chiasmic journey as do the trams in 'Aeolus'. They cross over and nearly meet on a number of occasions, only to be linked finally in the episode of extremes, 'Circe', where 'Jewgreek is greekjew. Extremes meet' (p. 474).[74] Though their meeting is, to an extent, anti-climactic, the criss-crossing of their identities continues in 'Ithaca' where their two names, Bloom and Stephen, transmute into the textual intersection of 'Bloom Stoom' and 'Stephen Blephen' (p. 635).

A slightly different use of chiasmus obtains in the relationship between the novel's characters and the social space of Dublin. The urban environment profoundly affects both the psychic space of the interior monologue and the somatic space of characters in the book. Joyce's interest in depicting the human body in the novel is well documented. Fourteen of its episodes are assigned bodily organs. This perhaps obscures the fact that in many ways Dublin is treated as an enormous body in this fiction: Joyce's 'epic of the human body'[75] is simultaneously an epic of the body of a city, as well as an exploration of the problematical relationship between corporeality and city streets.[76] In *Ulysses*, said Joyce, 'the body lives in and moves through space and is the home of a full human personality'.[77] Such a claim suggests a fascinating congruence between body and space, viewing the body as a spatial entity and, crucially, tracing its movements through space itself. In this respect Joyce's views concur with those of Lefebvre, who assigns a key role to the body in his theory of social space: 'it is by means of the body that space is perceived, lived – and produced' (*PS*, p. 162). The focus in *Ulysses* upon the body, especially that of Bloom, is one of its most innovative features, exemplifying Lefebvre's claim that the 'body, at the heart of space and of the discourse of power, is irreducible and subversive'.[78]

This interaction of space and body is clearly seen in 'Lestrygonians', the episode concerned with Bloom's tramp around the city. As he crosses the River Liffey, Bloom realises he is hungry and starts to think of where to take his lunch. Food dominates the symbology of this episode, with Joyce describing the style as 'peristaltic prose'. The Gorgonzola sandwich and glass of burgundy that Bloom consumes in Davy Byrne's pub also develop the chiasmus employed in 'Aeolus'. Eating brings that which is exterior into the body, beginning the body's process of digestion and excretion; eating is a form of chiasmus, something that criss-crosses the body, entering in, only to exit later.

'Lestrygonians' also explores another form of chiasmus in the relation between the city and the interior monologue: if Bloom visually ingests what he sees around him, devouring the many signs and adverts of the Dublin streets, eventually the city bites back, and gobbles him up. As Umberto Eco notes, in *Ulysses* 'personal identity is questioned. In the flow of overlapping perceptions during Bloom's walk through Dublin, the boundaries between "inside" and "outside", between how Bloom endures Dublin and how Dublin acts on him, become very indistinct.'[79] The interiority of the monologue always partakes of the exteriority of the city and its inhabitants, creating a constant effect of chiasmus, crossing between inner and outer spaces.

This feature is most noticeable in 'Lestrygonians' when Bloom crosses the front of Trinity College and notices the incessant movement of the trams: 'Trams passed one another, ingoing, outgoing, clanging. Useless words. Things go on same; day after day; squads of police marching out, back: trams in, out' (p. 156). The pulsation of the trams is transferred into the more general principle of the flux of modern life, as Bloom thinks of the death of Dignam followed by the birth of Mina Purefoy's child. Bloom has ruminated earlier upon this principle of perpetual motion in an advert for Kino's trousers placed on a boat anchored on the Liffey: 'Good idea that. Wonder if he pays rent to the corporation. How can you own water? It's always flowing in a stream, never the same, which in the stream of life we trace. Because life is a stream. All kinds of places are good for ads' (p. 146). The exterior space of the adverts and the trams outside Trinity crosses into Bloom's consciousness, fuelling his next snapshot of the city, a dazzling description of urban life itself:

> Cityful passing away, other cityful coming, passing away too: other coming on, passing on. Houses, lines of houses, streets, miles of pavements, piledup bricks, stones. Changing hands. This owner, that ... Piled up in cities, worn away age after age. Pyramids in sand. Built on

> bread and onions. Slaves Chinese wall. Babylon. Big stones left. Round towers. Rest rubble, sprawling suburbs, jerrybuilt, Kerwan's mushroom houses, built of breeze. Shelter for the night.
>
> No one is anything.
>
> This is the very worst hour of the day. Vitality. Dull, gloomy: hate this hour. Feel as if I had been eaten and spewed. (p. 157)

This is what a city amounts to: not a static collection of buildings, but a constant process of building and rebuilding of 'piledup bricks', with the neologism capturing perfectly the constructive activity of both brick and word.[80] This 'stream of life' is the fluid modernist city that reduces interiorised identity to nothing; Bloom has already put down his inability to remember somebody's name to the '[n]oise of the trams probably' (p. 148). Now he is eaten and regurgitated by the city itself, his body demolished and rebuilt like a new suburb, turning Bloom quite materially into a Dubliner. Stephen's intellectual questioning of identity in the next episode, 'Scylla and Charybdis', is here pre-empted by a grossly material interrogation into the spaces – both bodily and urban – that constitute the very fibre of Bloom's identity.

The representational spaces of *Ulysses* constantly question identity, rendering the distinctions between body and city, inner and outer, somewhat meaningless. The spatial form of the novel, however, also draws upon such chiasmic movements, with the text itself caught up in a disorientating stream of coming and passing. Bloom's impression of being devoured by the city is also a good depiction of how the novel cannibalises its own space, turning words and signs into textual nibbles for future use. In this way chiasmus appears once again, as the novel quotes and requotes from its own pages, criss-crossing backwards and forwards across the body of the text. For example, as Bloom's 'slow feet walked him riverward' (p. 144) he is handed a throwaway, a leaflet advertising a religious meeting and the imminent arrival of Elijah. The throwaway is, of course, anything but, as a horse of the same name wins the Ascot Gold Cup that day, and Bloom has already inadvertently handed this tip to Bantam Lyons in the earlier 'Lotus Eaters' episode. That which is outside, the throwaway, crosses over into the head of Bloom, as do countless of the many advertisements he perceives during the day. But equally the text *throws away* nothing; *Ulysses* we might say, is extremely environmentally friendly,[81] as words and phrases are ingested, digested and excreted elsewhere.[82] The words of the text thus become, to return to another metaphor, like the 'piledup' bricks of the houses of Dublin pondered by Bloom, reconstructed anew by the author as master builder.

Chiasmus, then, is a rhetorical figure that operates by means of passing through space, whereby linguistic patterns are traced upon the page. It is also, as I have suggested here, a figure that can accurately describe many other processes at work in *Ulysses*, most noticeably the many movements between material and metaphorical space, such as that of Dublin and the body or the thoughts of a character such as Bloom.

Objects, places, forces: 'Wandering Rocks'

The most complex set of movements through modern spaces in *Ulysses* is found in 'Wandering Rocks'. In many ways this is the episode in which the social space of Dublin most visibly shapes the morphology of Joyce's narrative space. In the Linati schema the episode utilises 'Objects, Places, Forces', and it is this whirligig texture of city life that fills its pages. 'Wandering Rocks' consists of eighteen short scenes and a coda dealing with the meetings, travails and travels of a large number of the central, and marginal, characters encountered so far. These consecutive mini-episodes, however, swim against the dominant current of narrative time by means of some thirty-three interpolations that interrupt the flow of the prose and which could not have been witnessed by those present in the episode. Their effect is, firstly, to thoroughly disorientate the reader's understanding of the spatial form of the narrative, thus approximating to the ceaseless flurries of the city as noted by Bloom in 'Lestrygonians'. Secondly, the interpolations foreground the category of space itself due to the narrative presentation of different people simultaneously in locations throughout the city.[83]

In this episode it is the force of motion that is most prominent, whether this is by foot or by motor, bearing out Seidel's over-assessment that Joyce in *Ulysses* overall 'wants direction more than he wants rootedness'.[84] In de Certeau's terms this episode offers a plethora of spatial stories in a series of mini-tours through the city: it is up to the reader to supply the missing map.[85] In 'Wandering Rocks' Father Conmee travels by tram to the very outskirts of the city; Boylan arranges to have a selection of fruit sent by tram to the 'invalid' Molly; there's a bicycle race through the crowded streets; two cars of tourists pass across the front of Trinity College while Stephen talks in Italian to Almidano Artifoni; an ambulance car shoots through the streets; and we follow the 'throwaway' proclaiming the appearance of the prophet Elijah, jettisoned earlier by Bloom, slowly meandering eastwards along the River Liffey and out to sea. Mainly, however, it is walking that permeates the episode, and so much wandering goes on that it is hardly surprising that one minor character, Jimmy

Henry, complains of the corns on his feet (p. 237). Joyce also draws attention to the difficulties some bodies have in crossing through city space by having Father Conmee come across 'a onelegged sailor, swinging himself onward by lazy jerks of his crutches' (p. 210).[86]

One comic example of pedestrian motion in this episode is the sandwichboard men of HELYS stationery store, for which Bloom once worked. The HELYS men who 'plodge' the Dublin streets are yet another metaphor, taken from actual urban experience, for how the space of the text is encountered. They wander the city, back and forth, sometimes getting their letters out of sequence, as when the H is separated from the ELYS (p. 243), or the Y lags behind the rest in order to snatch some food in 'Lestrygonians' (p. 147). The text of HELYS wanders, not like rocks, but as a set of disconnected signifiers in search of a signified that will unite them into a coherent sign. The sandwichboard men illustrate de Certeau's discussion of walking in the city: 'To walk is to lack a place ... The moving about that the city multiplies and concentrates makes the city itself an immense social experience of lacking a place'(*PE*, p. 103). The HELYS men construct the city as a site of perpetual movement, with no sense of rooted place. They are a kind of spatial counter to the Dublin place-names, representing Joyce's text as an extended 'pedestrian speech act' and demonstrating that walking in the city is, as de Certeau argues, 'a process of *appropriation* of the topographical system on the part of the pedestrian' (*PE*, p. 97). Just as the men 'eeled' between Dublin streets, so Joyce's text wriggles around in this episode, refusing to construct a static 'identification of place'.

Joyce's avant-garde experiments with narrative time and space in 'Wandering Rocks' are far from merely exercises in style. In this itinerant episode narrative wandering is finally curtailed by a stark operation of power over social space. The interpolations and temporal simultaneities of 'Wandering Rocks' are intimately connected to the spatial politics of the city when the final coda of the episode follows the procession through the Irish capital of the British governor-general in a viceregal cavalcade. Once more, recalling Joyce's use of the Nelson pillar, the political relations between Ireland and Britain are figured by means of a spatial image: here, not a static monument, but the triumphal display of imperial power through the streets of the subjugated capital. The cavalcade is an example of what Lewis Mumford describes as 'baroque power', and it is significant that many of the Dublin subjects who have been tramping around the city here stop and salute the cavalcade.[87] Although much of 'Wandering Rocks' is dominated by the familiar urban trope of the labyrinth, the viceregal procession operates as a kind of centripetal force in this episode.[88] The myriad spaces of the episode, with its multiple characters and shifting perspectives, are unified by the all-embracing visibility of the viceregal carriages in the closing coda. Duffy suggests that the viceregal parade is insignificant and 'nothing but show';[89] but visual show is precisely the point here. The cavalcade operates as a kind of panopticon, in Foucault's conceptual sense of a form of power based upon surveillance and visibility: because the viceroy is seen by all, he too sees all, and spatially his gaze rules over all Dublin.[90] It is because of the spatial power represented here that John Wyse Nolan describes the governor-general as the 'general governor of Ireland' (*U*, p. 238).

That Joyce thought the viceregal procession important is shown not only by the fact that it enacts a narrative and a spatial closure to this episode, perhaps the most fluid of all the episodes in *Ulysses*, but because the event to which it refers – a money-raising charity bazaar for a Dublin hospital – did not happen on the day of the novel, 16 June 1904, but several days earlier and did not involve a viceregal cavalcade.[91] The procession is thus a false narrative interpolation – it could not have been seen in either the time or the space of the novel – but is another instance of Joyce's exploration of spatial politics. Other 'errors' in the account of the cavalcade testify to this aspect: the cavalcade crosses the '*Royal* Canal bridge' in the south-east of the city, where the procession should have encountered the *Grand* Canal. This only reinforces the imperial power over Dublin space symbolised by the cavalcade. Indeed, the course of the parade is itself of significance. It departs from the viceroy's official residence in Phoenix Park, with its monument to Anglo-Irish military hero, the Duke of Wellington. From the west of the city it travels along the north banks of the river past '*Kingsbridge*' and through many of the main thoroughfares of Dublin, crossing the Liffey at *Grattan*'s Bridge into *Parliament* Street and alongside the front of Trinity College, before driving off south-eastwards to the bazaar. Henry Grattan was leader of the last Irish parliament before the Act of Union, at the end of the eighteenth century, the site of which is the Bank of Ireland alongside Trinity College. Joyce thus quietly draws our attention to this imperial history by having the current ruler of Ireland, the governor-general, pass through sites redolent of past Irish parliamentary power, as well as toponyms with a suitably regal flavour.

The geographical wanderings of the procession demonstrates the technique employed by Joyce in this episode: that of a 'shifting labyrinth between two shores',[92] here not the Mediterranean, but those of Ireland and Britain. The cavalcade also crosses, unlike the characters in 'Wandering Rocks', from the north to the south bank of the River Liffey, emphasising the expansion of the governor's domain. The imperial power displayed in the viceroy's cavalcade forms a sharp contrast to the many Irish nationalist heroes noted in this episode, such as Emmett, Tone and Parnell. The viceregal procession thus passes the place where Robert Emmet was hanged (p. 230–1) and, indicating a particularly pregnant absence, not far from 'the slab where Wolfe Tone's statue was not' (p. 220). A foundation stone was laid in 1898, in a corner of St Stephen's Green, but no statue followed. The absence of a statue to this Irish nationalist leader contrasts sharply with the presence of the British viceroy on the Dublin streets and the monuments to military heroes such as Nelson and Wellington.

Vincent J. Cheng has indicated the political significance of the viceroy's form of transport through the Dublin environs. The horse, Cheng argues, 'has been a traditional symbol of power ... and denotes the rule of authority' in Ireland, signified by the image of British cavalry and the fact that under the Penal Laws no Catholic could own a horse over the value of £5.[93] Not surprisingly, notes Cheng, horses are one of the things that Stephen says he is afraid of in *A Portrait of the Artist as a Young Man*.[94] In 'Wandering Rocks', argues Cheng, 'the Viceroy's horses became a metonymy for the power of Empire'.[95] Significantly, the cavalcade passes the equestrian statue of King Billy opposite Trinity College, a symbol of Protestant victory over Catholicism at the Battle of the Boyne, and removed in 1929 after the creation of the Irish Republic. The 'foreleg of King Billy's horse pawed the air' (p. 243) as the viceroy passes, as if in salute. The supremacy over Dublin space embodied in the horse-drawn procession is also indicated by the fact that 'the tram and Spring's big yellow furniture van had to stop ... on account of it being the lord lieutenant' (p. 243). The whirl of urban life in 'Wandering Rocks' is thus subject to a constraining force that pulls this episode to a formal close. Textual space is thus closed down by the power of the viceroy's hold over the social space of the city.

Resistance to that power is, however, evident, for although certain Dublin citizens (including Stephen's father) acknowledge the viceroy, and earlier in the day Bloom had considered going to the charity bazaar (p. 174), others refuse to salute his presence. The viceroy is 'unsaluted by Mr Dudley White'; the River Poddle offers 'in fealty a tongue of liquid sewage' (p. 242) and John Wyse Nolan 'smiled with unseen coldness towards the lord lieutenantgeneral' (p. 243). Pointedly, Parnall's brother refuses to look up from his chessboard in the D.B.C. teashop. The final acknowledgement of the viceroy comes from the 'sturdy trousers' of Almidano Artifoni (p. 244), an absurd image that comically subverts the relations of power between governor and governed.

It is worth emphasising again how the spatial form of an episode such as 'Wandering Rocks', with its stress upon the simultaneity of characters in time and space, is not just an impressive piece of experimental fiction. Kern, for instance, notes that the cavalcade's route through the city unites all the earlier incidents in the episode, in order to 'suggest the spatial interrelatedness of the city ... Although the cavalcade moves sequentially, the glimpses that some of the characters had of it in earlier sections anticipate the final summation of the simultaneity of its movements.'[96] This is true enough as a stylistic description, but ignores the political implications of Joyce's transformation of social space into textual space. The cavalcade's linear journey is a summation, but one which connotes a panoptic perusal over the surrounding streets, a force signifying a colonial 'representation of space'. The spatial form of 'Wandering Rocks' within *Ulysses* also partakes of the spatial politics explored in the book. The episode is positioned centrally in the novel, and its eighteen scenes formally parallel the eighteen episodes of the novel overall (though with little corollary in terms of their content). This pattern leaves the coda of the viceregal cavalcade at the end of the episode as lacking in a formal parallel to any chapter in the book: it is as if the coda exists outside of the space of the text itself. As the cavalcade forcefully moves through the 'shifting labyrinth between two shores' it seems to exist somewhat heterotopically, in a kind of non-place. However, this heterotopic spatial power parallels the way that *Ulysses* is structurally contained (or constrained) between two other sites of colonial power. The novel starts in the Martello tower and ends with Molly remembering her childhood in Gibraltar, site of a British military garrison and, according to one late nineteenth-century guidebook used by Joyce, 'the centre of the military life of England'.[97] Narrative morphology thus matches the spatial politics of the text. 'Wandering Rocks' once again demonstrates how the metaphorical space of its experimental pages must be understood in relation to the material spaces of Empire.

Watch! Watch! Sexual spaces

The nature of the power exemplified by the viceroy's triumphal procession through Dublin relies upon visibility: while he gazes upon his imperial subjects, they gaze, and sometimes salute, his presence. Joyce thus indicates how the interplay of looking in the city produces an important set of power relations. Another form of this investigation into the politics of the gaze in *Ulysses* concerns the gendered and sexualised nature of social space in the city. Feminist critics have often interrogated Joyce's work for his representation of women and men, but few have noted how precisely such depictions locate gender and sexuality within the social spaces of the modernist city.[98]

Raymond Williams argued that one important way in which modernism negotiated the city was by the development of a 'perceptual subjectivity'.[99] Social anomie and discontinuity were transformed into a problematical visual perception of others, since full human relationships were no longer readily available. More recent feminist criticism has shown how gender and sexuality help organise this 'perceptual subjectivity'. In chapter 3 I drew attention to the ways that men and women look and are looked at in the city, demonstrating how in Imagist poems social space is connected to sexualised and gendered spaces. Elizabeth Wilson's discussion of the *flâneur* sheds light on Bloom's experience of the city, when she suggests that the *flâneur* is a forlornly fictional creature, representing the threat of urban anonymity. Bloom as *flâneur* can be seen as a 'shifting projection of angst', using the 'male gaze' as a defensive strategy to cope with the vortex of the modern metropolis and the fear of symbolic castration by the female in the city. The tormented Bloom of 'Lestrygonians', consumed and spat out by the city, responds to such treatment in a series of compensatory libidinal gazes that try to demonstrate his visual power over women, but which ultimately shows him being undone by the urban experience.

Bloom's heightened visual sense is partly conditioned by his profession – he always has an eye open for where in the city a good advertisement might be placed. More than just a part of his working life, however, Bloom is always alert to the many pleasing visual images of the modern city. As he waits to purchase his breakfast kidney in Dlugacz's pork butchers in 'Calypso', Bloom notices the daughter of a neighbour. His thoughts drift between an image of her strongly beating a carpet on the clothesline (which hints at Bloom's masochistic desires), the contents of the butchers, and an advert for a model farm in a newspaper sheet used by the butcher to wrap the food. As the young woman leaves the shop Bloom's various visual images coalesce around the idea of following her down the street: 'To catch up and walk behind her if she went slowly, behind her moving hams. Pleasant to see first thing in the morning' (p. 57). But the woman escapes being reduced to another pork product for Bloom's visual feast and disappears down the street. As Bloom takes his kidney, the butcher slyly comments to Bloom, 'Another time' (p. 58), indicating a male heterosexual bond between them around voyeuristic pleasures. At a number of points throughout the day, Blooms seems to be searching for a replacement woman to satisfy his missed scopic breakfast.

In 'The Lotus Eaters' Bloom's peregrinations across the city are halted outside the Post Office by a chance meeting with McCoy. While McCoy tries to engage Bloom in conversation, Bloom's attention is directed to a woman about to board a cab in front of the Grosvenor Hotel. Bloom's thoughts signal his fetishistic desire, as he inspects her '[h]igh brown boots with laces dangling', and is excited by the prospect: 'Proud: rich: silk stockings.' But his visual thrills are thwarted by the appearance of a tram:

> Watch! Watch! Silk flash rich stockings white. Watch!
> A heavy tramcar honking its gong slewed between.
> Lost it. (p. 71)

Bloom is annoyed, and curses the 'noisy pugnose' of the tram, feeling suddenly 'locked out' of the sexual moment. The interruption of Bloom's pitiful glimpse of stocking indicates how the masculine gaze is foiled by this symbol of modern metropolitan life. The opportunity to fix libidinally the image of the anonymous woman is countered by the swift, and unexpected, movement of the tram. Urban flux thus disturbs the quest for stability represented by Bloom's visual desire. We later learn that in his 'middle youth' Bloom often sat watching, through the window of his house, 'the spectacle offered with continual changes of the thoroughfare without, pedestrians, quadrupeds, velocipedes, vehicles, passing slowly, quickly, evenly, round and round and round the rim of a round precipitous globe' (p. 633). Such images of the vertiginous city are, however, viewed from indoors, safe from the bustle of the traffic. When older, and outside in the swirling spaces of the city proper, Bloom's ocular obsessions are often transferred to matters of sexual pleasure, as in the woman outside the Grosvenor; trams, however, now represent an altogether more threatening spectacle.

Later, in 'Nausicca', Bloom obtains a more physical satisfaction for his erotic visions when he masturbates upon the strand while watching Gerty MacDowell. This consummation of his gaze prompts Bloom to recall his earlier scopic *interruptus* outside the Grosvenor Hotel: 'Made up for that tramdriver this morning' (p. 351) is his perky comment after his emission. It is striking, however, that this voyeuristic fulfilment occurs in a quite separate space, outside of the boundary of the city, safe on the Sandymount shoreline where no tram can disturb Bloom's business.

As with the rest of the novel, in 'Circe' Bloom's encounter with the tram is recycled once more. As he walks through nighttown Bloom narrowly avoids colliding with a tram, an incident of some importance for his sexual identity: 'He looks round, darts forward suddenly. Through rising fog a dragon sandstrewer, travelling at caution, slews heavily down upon him, its huge red headlight winking, its trolley hissing on the wire. The motorman bangs his footgong (p. 414). As Bloom 'blunders stifflegged' out of the way, the tram brakes heavily and the 'pugnosed' driver harangues Bloom, who thinks the driver might 'be the fellow balked me this morning with that horsey woman' (p. 415). This incident is a kind of reproach to Bloom's onanistic gaze at Gerty. In the morning the tram interrupted only Bloom's gawping, cutting through the spatial distance required for the fuelling of desire between the male voyeur and the female image.[100] But after Bloom's desire is satiated the tram returns, symbolically scolding him for his pleasures. 'Circe' is the episode of inversions, where Bloom's gender identity is upturned into the new 'womanly man', and the threat posed to the self-possession of the male gaze by the tram is perhaps the start of this unsettling process. After avoiding the tram Bloom is reminded of the pain that has appeared in his stomach, which then becomes accompanied by a headache. This he attributes to 'Monthly or effect of the other. Brainfogfag. That tired feeling. Too much for me now. Ow!' (p. 415) Again his masculinity is undermined by his worry that his 'brainfogfag' and lack of concentration is due to a menstrual period. The fog here seeping into his head rises from the river – 'Snakes of river fog creep slowly' (p. 412) – and demonstrates once again how the exterior urban environment crosses over into the interior space of character in the novel. Urban flux here, though, prompts this disorientation of sexual identity. The tram represents the literal dangers of transport in congested urban spaces, shown when Bloom muses that he might have lost his life to that 'mangongwheeltracktrolleyglarejuggernaut' (p. 429). This compound neologism is a monster word that typographically matches the hugeness of the threat from the tram. But the tram also operates symbolically, challenging Bloom's masculinity as a prelude to its destruction in the brothel. Bloom's aloof gaze is deeply compromised here, and bears out Wilson's point that the *flâneur* 'embodies the Oedipal under threat. The male gaze has failed to annihilate the castrate, woman', and so is pitched into the hurly-burly of metropolitan life.[101]

Ulysses seems an exemplary text to demonstrate Edward Said's claim that contests over geography involve not only the violence of 'soldiers and cannons' but also 'forms ... images and imaginings'.[102] *Ulysses* draws upon material spaces and places, their social character transformed by the geographical imagination of the writer. Nelson's pillar, standing proudly at the start of 'Aeolus', indicates a specific geographical fight over the lived meanings of city space. Joyce's representational space contests the official conceptions of space, indicating the complex imperial history of Dublin's streets. There is, as Duffy suggests, no nostalgic discourse of place as fixity operating in this text: all particular locations are excavated for the spatial stories they reveal of the history of Dublin.[103] Repeatedly Joyce draws attention to the political dimensions of city space in his novel, from the colonial cavalcade to the sexualised gaze of Bloom the voyeur.

In *Ulysses* social space is shown to be, in Lefebvre's terms, a hypercomplex phenomenon, where boundaries between different geographies are constantly being traversed. The interior space of consciousness is always infused with the exterior social spaces of Dublin, demonstrating the profound intermingling between material and metaphorical spaces. Joyce's employment of the rhetorical figure of chiasmus is a textual reminder of this movement between diverse spatial regimes. Indeed, the feeling that 'rapid motion through space elates one' aptly characterises the experience of reading *Ulysses*, where we Dart[104] from one space to another much quicker than any Dublin tram. Joyce's formal experimentation shows a modernism that has plunged into the flux and 'civilisation of luggage' that earlier writers found so unnerving. *Ulysses*, then, offers us a textual space that, unlike Imagism and Forster, unequivocally celebrates a movement through modernity. For Joyce, as for de Certeau, the tour always precedes the map. However, while celebrating the multiple tours through the geography of Dublin *Ulysses* reveals an experience of modernity that can only fully be understood in terms of Ireland's colonial spatial history.

Notes

1 For further information on the race see Donald T. Torchiana, *Backgrounds for Joyce's 'Dubliners'* (Boston, MA, Allen & Unwin, 1986), pp. 77–90. Joyce interviewed one of the French drivers in the race for the *Irish Times*: see James Joyce, 'The motor derby' in James Joyce, *Occasional, Critical, and Political Writing*, ed. Kevin Barry (Oxford, Oxford University Press, 2000).
2 James Joyce, *Dubliners* (London, Penguin, 1992), p. 35.
3 Letter to C. P. Curran, July 1904, *Letters of James Joyce*, ed. Stuart Gilbet (New York, Viking, 1957), vol. 1, p. 55.
4 Joyce, *Dubliners*, p. 37.
5 Joyce, *Dubiners*, p. 38.
6 See Michael North, *Reading 1922: A Return to the Scene of the Modern* (Oxford, Oxford University Press, 1999), for a cultural history of this year.
7 For postcolonial studies of Joyce see, among others, Vincent Cheng, *Joyce, Race and Empire* (Cambridge, Cambridge University Press, 1995); Emer Nolan, *James Joyce and Nationalism* (London and New York, Routledge, 1995); and Enda Duffy, *The Subaltern 'Ulysses'* (Minneapolis, Univeristy of Minnesota Press, 1994).
8 David Harvey, *The Condition of Postmodernity*, p. 273.
9 Joyce, in conversation with Arthur Power, quoted in Ellmann, *James Joyce*, p. 505.

10 Doreen Massey, 'The conceptualisation of place' in Doreen Massey and Pat Jess, eds, *A Place in the World? Places, Cultures and Globalization* (Oxford, Oxford University Press, 1995), p. 69.

11 James Joyce, *Ulysses*, ed. Jeri Johnson (Oxford, Oxford University Press, 1993).

12 See, for example, Kern's argument that the 'real action takes place in a plurality of spaces, *in a consciousness* that leaps about the universe and mixes here and there in defiance of the ordered diagramming of cartographers' (my emphasis): Stephen Kern *The Culture of Time and Space 1880–1918* (Cambridge, MA, Harvard University Press, 1983), p. 149.

13 See Michael Seidel's excellent *Epic Geography: James Joyce's Ulysses* (Princeton, NJ, Princeton University Press, 1976) which does not consider the political ramifications of geography in the novel. Two recent essays which take an approach to space and geography in Joyce similar to mine are Enda Duffy, 'Disappearing Dublin: *Ulysses*, postcoloniality and the politics of space', and Marjorie Howes, '"Goodbye Ireland I'm going to Gort": geography, scale and narrating the nation', in Derek Attridge and Marjorie Howes, eds, *Semicolonial Joyce* (Cambridge, Cambridge University Press, 2000). In particular Duffy, in a rich and nuanced argument, suggests that *Ulysses* is informed by a spatial politics that rejects a discourse of place, and which can help understand the colonial status of the city. Interestingly, he also links the unseen spaces of Dublin to Joyce's sense of a future Ireland freed from a constraining myth of national territory. While agreeing with the general approach of Duffy, particularly his use of Lefebvre, I disagree with certain elements of his reading of the text.

14 This view of the 'Hades' chapter was suggested by John McIntyre's seminar paper 'Confining plots: modernist space in Joyce's "Hades"', delivered at the 'Modernism and the city' seminar at the New Modernisms 2 Conference, University of Pennsylvania, October 2000.

15 Joyce, quoted in Frank Budgen, *James Joyce and the Making of 'Ulysses'* (1934) (Bloomington, Indiana University Press, 1960), pp. 67–8.

16 'After the Race', *Dubliners*, p. 39.

17 Michael Groden, *'Ulysses' in Progress* (Princeton, NJ, Princeton University Press, 1977), p. 60.

18 Letter to Stanislaus Joyce, c. 24 September 1905; quoted in Richard Ellmann, *James Joyce*, revised edn (Oxford, Oxford University Press, 1983), pp. 163, 208.

19 On newspapers in this chapter see Cheryl Herr, *Joyce's Anatomy of Culture* (Urbana, University of Illinois Press, 1986), ch. 2.

20 See Groden, *'Ulysses' in Progress*, p. 110.

21 Keating, for example, reads this opening as an attempt to render the universal characteristics of the modern metropolis, with a kind of Dickensian eye for detail: see Peter Keating, 'Literature and the metropolis', in Anthony Sutcliffe, ed., *Metropolis 1890–1940* (London, Mansell, 1984).

22 See Gary Granville, ed. *Divided City: Portrait of Dublin 1913* (Dublin, O'Brien Educational, 1978), p. 69.

23 Thus I disagree with Duffy's assertion that official monuments are given 'short shrift' in *Ulysses*; see Duffy, 'Disappearing Dublin', p. 49.

24 For information on Wilkins see R. W. Liscombe, *William Wilkins 1778–1839* (Cambridge, Cambridge University Press, 1980).

25 Samuel A. Ossory Fitzpatrick, *Dublin: A Historical and Topographical Account of the City* (1907) (Dublin, Tower Books, 1977), p. 298.

26 Ossory Fitzpatrick, *Dublin*, p. 299.

27 Jacques Derrida, *Writing and Difference*, trans. Alan Bass (London: Routledge, 1978), p. 279.

28 Luke Gibbons, *Transformations in Irish Culture* (Cork, Cork University Press, 1996), p. 145.

29 Richard O'Brien, *Dublin Castle and the Irish People* (Dublin, Gill; London, Kegan Paul, Trench & Trubner, 1909); quoted in Herr, *Joyce's Anatomy of Culture*, p. 86.

30 Nelson thus resembles a kind of imperial panopticon; see Michel Foucault, *Discipline and Punish: The Birth of the Prison* (Harmondsworth, Penguin, 1977).

31 An interesting comparison can be drawn between this assault upon a prominent city monument and the destruction of the Vendôme Column in Paris by the Communards, as discussed by Kristin Ross, *The Emergence of Social Space: Rimbaud and the Paris Commune* (London, Macmillan, 1988), pp. 5–8. As Ross notes: 'An awareness of social space, as the example of the Vendôme Column makes clear, always entails an encounter with history – or better, a choice of histories' (p. 8).

32 *Irish Times*, 8 March 1966.

33 See Patrick Henchy, 'Nelson's pillar', *Dublin Historical Record* 10: 2 (June–August 1948), 53.

34 Henchy, 'Nelson's pillar', p. 53.

35 *Irish Magazine*, September 1809; quoted in Henchy, 'Nelson's pillar', p. 59–60.

36 Quoted in Henchy, 'Nelson's pillar', p. 62.

37 *Irish Times*, 8 March 1966. As Emmett's famous last words get several mentions in *Ulysses* this resignification of the pillar might well have appealed to Joyce.

38 The paragraph on the Post Office was another addition to the 1921 revisions of the chapter.

39 In discussing Joyce's attitude in his critical writings to the physical force tradition in Irish nationalism, Emer Nolan suggests that this 'ambivalence' is part of Joyce's colonial identity: 'what his critics see as detached ambivalence was experienced as painful deadlock by Joyce. His writings about Ireland may not provide a coherent critique of either colonised or colonialist; but their very ambiguities and hesitations testify to the uncertain, divided consciousness of the colonial subject, which he is unable to articulate in its full complexity outside his fiction': Emer Nolan, *James Joyce and Nationalism* (London, Routledge, 1995), p. 130. The conflictual

textual space of 'Aeolus' is thus a good example of Joyce's divided vision of nationalism.

40 Harry Blamires, *The New Bloomsday Book* (London and New York, Routledge, 1988), p. 55. Moses has been spatially misplaced once already in the chapter, when J. J. O'Molloy wrongly refers to the statue of Moses by Michelangelo as being in Vatican City rather than Rome itself (p. 134). This error only helps reinforce the theme of imperial relations, however, since the Vatican is officially a capital within the capital city of another country, much as Dublin's 'capital' status is eccentric to the real capital of the British Empire.

41 The stress upon Nelson as an adulterer also chimes with the theme of Bloom's possible affair with Martha Clifford, and Molly's adulterous behaviour with Blazes Boylan. Yeats's poem 'The Three Monuments' slyly mocked the fact that Dublin's central street contained statues to three famous adulterers (Nelson, Parnell, and O'Connell).

42 Seidel, *Epic Geography*, p. 124.

43 For a dramatic representation of this point in relation to the ordnance survey of Ireland see Brian Friel, *Translations* (London, Faber, 1981). For another discussion of the politics of language in Ireland see Tom Paulin, 'A new look at the language question', in Field Day Theatre Company, eds, *Ireland's Field Day* (London, Hutchinson, 1985).

44 Budgen, *James Joyce and the Making of 'Ulysses'*, p. 152.

45 For the proleptic character of Joyce's text see Declan Kiberd, *Inventing Ireland: The Literature of the Modern Nation* (London, Vintage, 1996), ch. 19.

46 David Hayman suggested that the headlines are the work of 'something between a persona and a function', a body he terms the Arranger; see David Hayman, *'Ulysses': The Mechanics of Meaning* (Madison, University of Wisconsin Press, 1982), p. 122.

47 See Seamus Deane's argument that in *Ulysses* Joyce is concerned to linguistically 'subvert a political conquest'; that is, 'to make the world conform to words is a characteristic aspiration of a culture which has found it for so long impossible to makes its words conform to the world': Deane 'Joyce the Irishman', in Derek Attridge, ed., *The Cambridge Companion to James Joyce* (Cambridge, Cambridge University Press, 1990), p. 43.

48 For an overview of Irish trams see Jim Kilroy, '"Galleons of the streets": Irish trams', *History Ireland* 5: 2 (summer 1997), 42–7. Hugh Kenner, in *The Mechanic Muse* (Oxford, Oxford University Press, 1987), claims that the Dublin system was in 1904 'the world's most extensive' (p. 11); Kilroy rather more modestly estimates it as the seventh largest in the world.

49 Budgen, *Joyce and the Making of 'Ulysses'*, p. 129.

50 Ossory Fitzpatrick, *Dublin*, p. 341.

51 Ossory Fitzpatrick, *Dublin*, p. 343.

52 See Luke Gibbons, 'Montage, modernism and the city', *Irish Review* 10 (1991), 1–6, for a criticism of Franco Moretti's reading of 'backward' Ireland in *Ulysses*.

53 Luke Gibbons, *Transformations in Irish Culture*, Critical Conditions: Field Day Essays (Cork, Cork University Press, 1996), p. 6.

54 Seamus Deane, 'Dead ends: Joyce's finest moments', in Attridge and Howes, eds, *Semicolonial Joyce*, p. 26.

55 Budgen, *Joyce and the Making of 'Ulysses'*, p. 93.

56 See Mary E. Daly, *Dublin: The Deposed Capital. A Social and Economic History 1860–1914* (Cork, Cork University Press, 1984), pp. 171–4.

57 Both fares are for a penny, and thus were probably for short distances only, penny fares being introduced for such journeys in 1884. See Daly, *Dublin*, p. 174.

58 Seidel interestingly links the woman who enters the Martello tower with the two women who climb the towering Nelson's pillar in 'Aeolus'; see Seidel, *Epic Geography*, p. 141.

59 Mikhail Bakhtin, *The Dialogic Imagination: Four Essays*, ed. Michael Holquist, trans. Caryl Emerson and Michael Holquist (Austin, University of Texas Press, 1981), p. 252.

60 Howes, 'Geography, scale and narrating the nation', in Attridge and Howes, eds, *Semicolonial Joyce*, p. 63.

61 Mary E.Daly, *Dublin*, p. 54; for information on the cattle trade see pp. 8–9.

62 Fredric Jameson, *Modernism and Imperialism* (Derry, Field Day, 1988), p. 22.

63 Gerry Smyth, *The Novel and the Nation: Studies in the New Irish Fiction* (London, Pluto, 1997), p. 58. For a further consideration of the relation between city and country see Fintan O'Toole, 'Going west: the country versus the city in Irish writing', *The Crane Bag* 9: 2 (1985), 111–16.

64 See W. B. Yeats, 'The Fisherman' in *Yeats's Poems*, ed. A. Norman Jeffares (London, Macmillan, 1989), p. 251.

65 Smyth, *Novel and Nation*, p. 60.

66 See R. F. Foster, *Modern Ireland 1600–1972* (London, Penguin, 1988), p. 350.

67 The failed French invasion of the west of Ireland in 1798 is an incident referred to often in the pages of *Ulysses*.

68 The phrase is Wyndham Lewis's for Joyce's method; see 'An analysis of the mind of James Joyce' in *Time and Western Man* (1927) (Boston, MA, Beacon Press, 1957), p. 92.

69 Another geographical parallel noted by Seidel is how the direction of the trams across the city apes the journey of Odysseus across the Mediterranean: Seidel, *Epic Geography*, p. 164.

70 Kenner, *Mechanic Muse*, p. 77.

71 Kenner, *Mechanic Muse*, p. 11.

72 Stuart Gilbert lists ninety-five rhetorical tropes in this episode; see his *James Joyce's 'Ulysses'* (1930; New York, Vintage, 1955), pp. 194–8.

73 A detailed reading of chiasmus and rhetorical parallelism is found in Murray McArthur, 'Rose of Castille/rows of cast steel: figural parallelism in *Ulysses*', *James Joyce Quarterly* 24: 4 (summer 1987), 411–22. Chiasmus in 'Aeolus' is

also discussed in John Paul Riquelme, *Teller and Tale in Joyce's Fiction* (Baltimore, MD, Johns Hopkins University Press, 1983), pp. 187–9.

74 This meeting refers to the Semitic Bloom and the supposedly Hellenic Stephen.

75 Budgen, *Joyce and the Making of 'Ulysses'*, p. 21.

76 Richard Sennett, in *Flesh and Stone: The Body and the City in Western Civilization* (London, Faber, 1994), argues that one of the key problems facing many contemporary Western societies is the 'sensory deprivation' and isolation of human bodies due to the spatial relations of urban life: the 'geography of the modern city', writes Sennett, 'brings to the fore deepseated problems in Western civilization in imagining spaces for the human body which might make human bodies aware of one another' (p. 21). For a psychoanalytically influenced account of these issues see Steve Pile, *The Body and the City: Psychoanalysis, Space and Subjectivity* (London and New York, Routledge, 1996).

77 Cited in Budgen, *Joyce and the Making of 'Ulysses'*, p. 21.

78 Henri Lefebvre, *The Survival of Capitalism* (1976); quoted in Derek Gregory, *Geographical Imaginations* (Oxford, Blackwell, 1994), p. 159.

79 Eco, quoted in Franco Moretti, *Signs Taken for Wonders: Essays on the Sociology of Literary Forms*, trans. Susan Fischer, David Forgacs and David Miller (London, Verso, 1983), p. 194.

80 Daly notes that in the closing years of the nineteenth century and the early 1900s Dublin was 'marked by a high level of building activity' (*Dublin*, p. 63).

81 This suggestion I owe to Barry Fegan.

82 Much of this added texture of quotation and recycled quotation in the novel was incorporated in what Groden terms the middle phase of the composition of the novel.

83 Kern, *Culture of Time and Space* (p. 77), interprets the emphasis upon simultaneity as being inspired by the cinema.

84 Seidel, *Epic Geography*, p. 247.

85 This strategy, suggests Enda Duffy, 'Disappearing Dublin' (pp. 51–2), turns the reader into a kind of colonial tourist to the city.

86 In a savage comment on the 'superiority' of Father Conmee, Joyce has him refuse to put money into the purse of the begging sailor because he believes the sailor to already possess a crown. This is contrasted with the generosity to the sailor of the fairly irreligious Molly Bloom later in the chapter (p. 216).

87 Mumford notes how street planners in the eighteenth century widened avenues in various European cities in order for the wealthy to parade along them in carriages: Lewis Mumford, *The City in History: Its Origins, its Transformations and its Prospects* (Harmondsworth, Penguin, 1966), p. 424.

88 See Bakhtin, 'Discourse in the novel', *Dialogic Imagination*, pp. 270–1, for the notion of the centripetal forces that strive to unify and centralise language amid the dispersing effects of linguistic heteroglossia.

89 Duffy, 'Disappearing Dublin', p. 49.

90 Foucault, *Discipline and Punish*, pp. 195–228.

91 According to the *Irish Times* for June 1904, the Mirus Bazaar, raising money for the Mercer's Hospital, ran from 31 May to 4 June, and was so popular it was extended for a further two days. On 16 June 1904 the viceroy was enjoying a stay in Lahinch, County Clare, at the Golf Links Hotel.

92 See the Linati schema for this description.

93 Cheng, *Joyce, Race and Empire*, p. 255.

94 *Joyce, Race and Empire*, p. 257.

95 *Joyce, Race and Empire*, p. 257.

96 Kern, *The Culture of Time and Space* , p. 77.

97 Susan Bazargan suggests that Joyce's choice of Gibraltar for Molly's youthful home allowed him to 'synecdochically represent British imperialism [and] project colonialism in its contemporary, international guise in his study of the displaced and the dispossessed', drawing a 'portrait of modern female colonial identity': Susan Bazargan, 'Mapping Gibraltar: colonialism, time, and narrative in "Penelope"' in *Molly Blooms: A Polylogue on 'Penelope' and Cultural Studies* (Madison and London, University of Wisconsin Press, 1994), p. 119. The guidebook used by Joyce is Henry M. Field, *Gibraltar* (1888); see Bazargan, p. 120.

98 Works which do consider the gendering of space in *Ulysses* and Joyce generally, but without employing an explicitly geographical approach, include Shari Benstock, 'City spaces and women's places in Joyce's Dublin', in Bernard Benstock, ed., *James Joyce: The Augmented Ninth* (New York, Syracuse University Press, 1988); Bonnie Kime Scott, 'James Joyce: a subversive geography of gender', in Paul Hyland and Neil Sammells, eds, *Irish Writing: Exile and Subversion* (Basingstoke, Macmillan, 1991); and Johanna X. K. Garvey, 'City limits: reading gender and urban spaces in *Ulysses*', *Twentieth Century Literature* 41: 1 (1995), 108–23.

99 Williams, *The Country and the City* (Oxford and New York, Oxford University Press, 1973), p. 245.

100 On the need for distance in the male gaze see Mary Ann Doane, 'Film and masquerade: theorizing the male spectator', *Screen* 23: 3–4 (September–October 1982), 74–87.

101 Elizabeth Wilson, 'The invisible *flâneur*', *New Left Review* 191 (January–February 1992), 109.

102 Edward Said, *Culture and Imperialism* (London: Vintage, 1994), p. 6.

103 Duffy, 'Disappearing Dublin', p. 48.

104 The Dart is of course the name for Dublin's present-day light transit system.

5 Virginia Woolf: literary geography and the kaleidoscope of travel

In her first review for the *Times Literary Supplement* in 1905 Virginia Woolf outlined her objections to one form of 'Literary Geography', as the title of her piece proclaimed. The review considered two books, *The Thackeray Country* and *The Dickens Country*, published in a series entitled 'Pilgrimage', which offered illustrated guides to the places represented in the work of various writers.[1] Woolf first discusses how certain writers, such as Scott or the Brontës, vividly evoke a country and its inhabitants. She then notes how the two titles under review are somewhat misleading, given the essentially cosmopolitan nature of Thackeray's writing and the London setting that predominates in Dickens. Woolf also dislikes the photographs in the books of places where fictional characters might have lived: 'To imprison these immortals between brick walls strikes one as an unnecessary act of violence; they have always tenanted their own houses in our brains, and we refuse to let them go elsewhere.'[2] This division between external spaces and internalised 'houses in the brain' forms the crux of Woolf's reticence about this form of literary geography:

> A writer's country is a territory within his own brain; and we run the risk of disillusionment if we try to turn such phantom cities into tangible brick and mortar. We know our way there without signposts or policeman, and we can greet passers-by without need of introduction. No city indeed is so real as this that we make for ourselves and people to our liking; and to insist that is has any counterpoint in the cities of the earth is to rob it of half its charm.[3]

This chapter argues that Woolf's mature fiction, from *Jacob's Room* (1922) through to the end of the 1920s, concentrating upon *Mrs Dalloway* (1925), is dominated by this division between inner and outer space, between 'houses in the brain' and the city of bricks and mortar.[4] Woolf appears to refine this early statement on literary geography, as her novels constantly play across the spatial borders of inner and outer, constructing a fiction that shows how material spaces rely upon imaginative conceptualisation, and how the territory of the mind is informed by an interaction with external spaces and places. In *Jacob's Room* Woolf pithily summarised these two spaces: 'The streets of London have their map; but our passions are uncharted'(*JR*, p. 82).[5] And contemplating how to construct *The Waves* in 1928, Woolf mused, 'what is my own position towards the inner & the outer? I think a kind of ease & dash are good; – yes: I think even externality is good; some combination of them ought to be possible.'[6] Woolf does not reject externally mapped spaces; rather she charts psychic life via her use of stream of consciousness, criss-crossing the liminal regions of inner and outer with 'ease and dash' to reveal how this division itself is somewhat false. Her work explores the 'ambiguous continuity' between different forms of hypercomplex space. Critics have analysed spatial aspects of Woolf's work, pointing to what Randall Stevenson terms a 'growing inclination to turn from the world to the mind', shifting attention to the 'disparities between the mind within and the world without'.[7] This approach, however, plays down the interaction between inner and outer spaces in Woolf's work, and how her texts continue to interrogate the external geographies of modernity. As Gillian Beer notes, in *Mrs Dalloway* Woolf 'sets out the topography of London as precisely as does Defoe' and the 'accounts of walks and of districts register the characters' social space as well as their separations'.[8]

Woolf is another writer fascinated by the idea of moving through the spaces of modernity, finding the material culture of transport to be a suitable 'vehicle' for rendering the quotidian experiences of the modern world. In another early review she expressed the wish that modern transportation should be celebrated in literature:

> Cultivated people grumble at trains, and, if they are old enough, prefer the days of the stage coach. When you crawled over the surface of the earth, and swayed in the ruts, and saw the whole landscape through the steam of four fine horses you knew it face to face, they argue, as one should know one's friends. But surely it is time that someone should sing the praises of express trains. Their comfort, to begin with, sets the mind free, and their speed is the speed of lyric poetry, inarticulate as yet, sweeping rhythm through the brain, regularly, like the wash of great waves.[9]

Again Woolf oscillates between external and internal space, with the speed of express trains transformed into the rhythms of the brain, before being likened to her familiar symbol, waves of the sea. Woolf views the trains not merely as something to be celebrated and described in a realist fashion, but as a speeding symbol of the experience of modernity itself; external reality collapses into internal space, only to be re-presented once again in a different outer space. In her early experimental short story 'The Mark on the Wall' she writes that 'if one wants to compare life to anything, one must liken it to being blown through the Tube at fifty miles an hour – landing at the other end without a single hairpin in one's hair!' and that it is this kind of experience that 'seems to express the rapidity of life, the perpetual waste and repair; all so casual, all so haphazard'.[10] Throughout her writings images of transport are employed in this analogous way for the sensations – the 'rapidity' – of modern life. James Donald and others have argued, convincingly, for the impact of cinema technology on Woolf's writing in the 1920s.[11] Part of this chapter explores the impact of another technology, that of transport, discernible in the theme of the interrogation of identity and in certain stylistic features of Woolf's writings. Different forms of transport, however, signify divergent facets of modernity for Woolf. The tube experience of 'perpetual waste and repair' seems a pessimistic one, whereas other forms of public transport, such as the bus, are represented in a more welcome fashion. From the late 1920s onwards the motorcar is also depicted positively by Woolf as an agent of spatial freedom. This chapter initially discusses Woolf's cartography of London in *Mrs Dalloway*, before focusing upon Woolf's representations of public and private transport.

London geographies

If Joyce was said to have written *Ulysses* with an open map of Dublin alongside him, it appears that Woolf also saw the potential of maps for the writing of fiction. Some of her reading notes on *The Canterbury Tales* from 1922 contain an outline map of Green Park, Stratton Street and Bond Street in preparation for Clarissa's early morning jaunt in the novel.[12] In an early review of Henry James's travel writings, Woolf enthused over the rich representational qualities of maps in contrast to written depictions of places:

> Indeed, it is safe to say that if you want to know the look of some town in Cornwall or Wales or Norfolk the best plan will be to get a map and study its portrait there. For some reason there is more of the character of a place in this sheet of coloured paper, with its hills of shaded chocolate, its seas of spotless blue, and its villages of dots and punctures than in all the words of an ordinary vocabulary, arrange them how you will. The swarm of names, the jagged edge of the coast-line, the curves that ships make ploughing round the world, are all romantic grains of fact brewed from the heart of the land itself, and sluggish must be the mind that would refuse to work with such tools as these.[13]

The visual texture of the map offers a fuller picture of the material space of landscape than any linguistic representation. *Mrs Dalloway* may be read as Woolf's attempt to overcome this perceived deficiency, by producing a textual space whose language accurately renders 'the heart of the land'.

Mrs Dalloway shares with *Ulysses* a highly developed geographical focus, a fact recognised by the many maps that accompany editions in the 1990s of some of Woolf's works.[14] Like Joyce, Woolf shows a concern for the dialectic of space and place, utilising the visual tool of the map to historicise spaces in her writing or, as she terms it, to draw 'romantic grains of fact brewed from the heart of the land itself'. *Mrs Dalloway*, therefore, shows how Woolf's depictions of place are always informed by a sense of social space, and of the deeper historical and political resonances of particular sites in the city. In de Certeau's terms, Woolf always offers us an actualisation of space rather than a mere identification of place.[15] 'Why do I dramatise London perpetually?'[16] asked Woolf late in her life. In attempting to answer that question numerous critics have explored her representations of the city. Feminist critics such as Squier have shown how Woolf's London writings develop from 'a position of anxious marginality' to 'an embrace of the political and aesthetic possibilities for a woman in the modern city'.[17] My approach, while indebted to such work, differs in its use of a critical literary geography.

Mrs Dalloway's locations are, interestingly, much more circumscribed than those in *Ulysses*. Dublin's borders, suburbs and hinterlands are fairly well travelled in *Ulysses*, while in *Mrs Dalloway* the representation of London is extremely limited. The novel is set mainly in Westminster and Mayfair, areas of power, wealth and influence, as would befit Clarissa Dalloway, wife of a Conservative MP. These are areas of traditional power in English society, and as such are somewhat distinct from Woolf's background in the liberal 'intellectual aristocracy' of the nineteenth century.[18] Even Bloomsbury, where Virginia and her husband Leonard lived, and the area from which their cultural formation takes its name, is not represented at length in the novel; only the lodgings of Peter Walsh and Septimus Smith are located in the Bloomsbury district. Thus the novel explores the areas of London most associated with traditional forms of aristocratic prestige, rather than with cultural power (Bloomsbury) or financial power (the City, towards which Elizabeth Dalloway peers, but does not go). Woolf is thus very acute in her use of representations of space to indicate power and status. But, where Forster in *Howards End* attempted to produce a condition of England novel that connects diverse spaces in city and country, *Mrs Dalloway*[19] does not really represent other parts of London. Unlike the tentative excursion in *Howards End* into the clerkly lands of Leonard Bast, south of the River Thames is not represented at all in *Mrs Dalloway*. Inner-city suburbs are mentioned, such as in the reference to the 'poor mothers of Pimlico' (*MD*, p. 19) waiting to see the queen outside Buckingham Palace; but Pimlico is itself close to the Houses of Parliament where Richard Dalloway works, and is not explored in the representational spaces of the novel. In a number of essays in the late 1920s and early 1930s Woolf described the pleasure she took in exploring London, including areas such as the East End's docks in 'The London scene'. In a diary entry for November 1923 Woolf daydreamed about wandering the city streets, 'walking say to Wapping',[20] another East End location. So the lack of a more socially panoramic picture of the city in *Mrs Dalloway* is, then, a clear indication of Clarissa Dalloway's own circumscribed social space.[21]

An interesting example of the literary geography of the novel occurs during a discussion of Septimus Smith's appearance, which might pass for that of 'a clerk, but of the better sort; for he wore brown boots' (*MD*, p. 75). Septimus, a war veteran, is said to be a 'border case' (*MD*, p. 75) between two niches in the clerkly class, a position that Woolf elucidates primarily by means of a cultural geography. Septimus 'might end with a house at Purley and a motor car, or continue renting apartments in back streets all his life' (*MD*, p. 75). Purley and the motorcar suggest the new Surrey suburbs, much promoted in the 1920s, and ownership of a car was, in the period, a prime marker of the newly affluent lower–middle class. Renting an apartment in a city-centre backstreet, however, indicates a failure to escape the metropolis for the suburbs. Interestingly, Septimus lives first in lodgings off the Euston Road (*MD*, p. 76), an area to the north of Bloomsbury and too close to the railway stations at Euston and St Pancras then to be considered desirable. After the war Septimus and his wife move to 'admirable lodgings off the Tottenham Court Road' (*MD*, p. 79), a slightly more respectable social location. It is presumably there, from out of the 'large Bloomsbury lodging-house window' (*MD*, p. 132), that Septimus leaps to commit suicide, indicating his rejection of, among other things, any aspiration towards the Purley suburbs.

The southern suburbs of London float through Woolf's work as a kind of negative heterotopia, non-places that can never enter into literary representation but which are gestured to as dismal destinations for certain lower middle-class characters. What Wyndham Lewis in 1914 referred to as the 'PURGATORY OF PUTNEY'[22] conveys what Woolf and her husband felt when they visited relatives in that suburb in 1920: 'The streets of villas make me more dismal than slums. Each has a cropped tree growing out of a square lifted from the pavement in front of it. Then the interiors …'.[23] This dislike of the middle-class tone of the suburbs is also registered in *Jacob's*

Room, with its gloomy images of 'the hordes crossing Waterloo Bridge to catch the non-stop to Surbiton', or the mason's van 'with newly lettered tombstones recording how some one loved some one who is buried at Putney.' (*JR*, p. 97) The same location is used later in the novel to refer to motorcars 'in which jaded men in white waistcoats lolled, on their way home to shrubberies and billiard-rooms in Putney and Wimbledon' (*JR*, p. 153). In her essay 'Street haunting' Woolf notes commuters who will be 'slung in long rattling trains, to some prim little villa in Barnes or Surbiton'.[24] These south London suburbs seem to represent both anonymity ('some one loved some one') and a kind of social and cultural death.

In contrast to the unenthusiastic associations that accrue to the suburbs, central London to the north of the Thames is described very positively in *Mrs Dalloway*. Early in 1915 Woolf lived with her husband in what was then the London suburb of Richmond, but was looking – without success – to rent a flat in the Bloomsbury area of the city. Woolf enthuses over the delights of the metropolis: 'I could wander about the dusky streets in Holborn & Bloomsbury for hours. The things one sees – & guesses at – the tumult & riot & busyness of it all – Crowded streets are the only places, too, that ever make me ... think.'[25] A few days later, Woolf decides she must take a trip to London 'for the sake of hearing the Strand roar, which I think one does want, after a day or two of Richmond ... one wants serious life sometimes'.[26] The external spaces of the metropolis were a crucial stimulus for Woolf's own psychic space, and it is their interaction that she would most fully explore in *Mrs Dalloway*.

It was only in 1924 that the Woolfs moved back to central London, to Tavistock Square, and started to enjoy walking through the streets as if 'upon a tawny coloured magic carpet'.[27] Much of Woolf's enthusiasm for London, albeit for slightly different areas, is found in the opening pages of *Mrs Dalloway*, when Clarissa 'plunges' into the streets of Westminster and crosses the busy thoroughfare of Victoria Street:

> For Heaven only knows why one loves it so, how one sees it so, making it up, building it round one, tumbling it, creating it every moment afresh; but the veriest frumps, the most dejected of miseries sitting on doorsteps (drink their downfall) do the same; can't be dealt with, she felt positive, by Acts of Parliament for that very reason: they love life. In people's eyes, in the swing, tramp, and trudge; in the bellow and the uproar; the carriages, motor cars, omnibuses, vans, sandwich men shuffling and swinging; brass bands; barrel organs; in the triumph and the jingle and the strange high singing of some aeroplane overhead was what she loved; life; London; this moment of June. (*MD*, p. 6)

Woolf captures the 'tumult & riot & busyness' of the city, and explores the perplexity of finding such a scene attractive. The city is understood as an ongoing process, what Forster diagnosed as the flux of modern life, and similar to Bloom's perception of Dublin in *Ulysses*. Instead of Forster's critique of the alienating city, Woolf associates the incessant buzz of urban life with an integral sense of human subjectivity. The city is seen, made and built around the human subject, who is engaged in 'creating it every moment afresh', showing a dialectical engagement of subject and object in the construction of the urban experience. Woolf slyly includes Clarissa's pious comment on the homeless as an indication of the limitations of her view of London life: Clarissa's delight in urban 'life' is shown to be restricted to the ability of her class to enjoy such freedoms. This eulogy to the city is thus quite aware of the political aspects of social space in the city, even as it simultaneously celebrates the thrill of urban existence.

Woolf captures the 'busyness' of the city in the spatial style of the final sentence of the extract above. It is a typically bravura display of Woolf's sentence construction, consisting of repetition ('in the') and a listing of objects and sights that is kept moving by the use of her favoured semi-colon. Alternating short and long clauses capture the motion and pauses of life in city streets. Though giving a predominantly visual description, commencing with 'people's eyes', Woolf is careful to include descriptions of urban sounds: 'the bellow and the uproar', 'the jingle' and the 'high singing' of the aeroplane. The sentence also includes phrases that seem to emanate from the city itself. Who, or what, is the object of 'the swing, tramp, and trudge'? Or of the 'the triumph and the jingle', which do not quite fit the description following of the plane? It is as if Woolf gives voice to the city itself, capturing the movements of urban modernity or what she called 'the texture of the ordinary day'.[28]

Not surprisingly, given the predominance of wave and sea symbolism in her writing, Woolf summarises this perception of the pleasures of city life as 'waves of that divine vitality which Clarissa loved' (*MD*, p. 8). But this positive metaphor for city life is questioned when Clarissa crosses St James's Park and pauses to look at 'the omnibuses in Piccadilly' (*MD*, p. 9). Now Clarissa's identity in relation to the city waves is made much more ambivalent, a perception prompted by looking at two types of motor vehicle, omnibuses and taxis:

> She felt very young; at the same time unspeakably aged. She sliced like a knife through everything; at the same time was outside, looking on. She had a perpetual sense, as she watched the taxicabs, of being out,

> out, far out to sea and alone; she always had the feeling that it was very, very dangerous to live even one day. (*MD*, p. 9)

Here the psychic identity of Clarissa is challenged by the material flux of the city traffic. The taxicabs symbolise isolated travel rather than the communal experience of train or bus travel, and show a different side to the 'life' and vitality of the city, where each day contains danger, particularly of urban anomie.

In a discussion of how urban life can be understood by the concept of the sublime, Elizabeth Wilson describes how the 'search for the meaning of the city, or for meaning in the city, takes many forms, not the least of which is to create new forms of beauty'.[29] That beauty, however, can be understood as sublime since it is always haunted by 'a kind of unease' at being overwhelmed by the immensity of the city. Clarissa herself, musing upon her lack of education or knowledge, notes how 'it was absolutely absorbing; all this; the cabs passing', an experience that leads her to doubt her own identity: 'she would not say of herself, I am this, I am that' (*MD*, p. 10). The city is not just fascinating to gaze at, but absorbing in the sense that Clarissa dissolves into the city, or, more precisely, into the movements of the city, here imaged by the passing taxis. This dissolution of identity is finally retrieved from the unease and danger Clarissa has just felt, and absorption into the city becomes strangely comforting:

> Did it matter then, she asked herself, walking towards Bond Street, did it matter that she must inevitably cease completely; all this must go on without her ... that somehow in the streets of London, on the ebb and flow of things, here, there, she survived, Peter survived, lived in each other, she part, she was positive, of the trees at home; of the house there, ugly, rambling all to bits and pieces as it was; part of people she had never met. (*MD*, p. 10)

The conventional urban fears of isolation and separation are thus overcome if the subject surrenders to the city. Here movement through modernity constitutes identity rather than representing a threat to it, as Forster felt in *Howards End*. Becoming is thus intrinsically linked to social space, rather than rejecting it for a Heideggerian place of dwelling. Also noticeable is a constant creative crossing of the boundaries between inner psychic space and outer social space. Clarissa Dalloway's positive immersion in the city justifies Woolf's own claim that only in the city could she think; singular identity passes away, just as the taxis passed through the streets, and Clarissa now exists within the flux of the cityscape. Peter Walsh, thinking of Clarissa a little later, claims that women 'attach themselves to places' (*MD*, p. 51); but her plunge into the city is more about carving out an identity by passing through space.

In contrast to Clarissa's ecstatic trip into the city, Peter Walsh's engagement with London reveals a quite different political geography of the city, one utilising gender categories to organise its representational spaces. While Clarissa strolls through the commercial streets of Bond Street and Piccadilly, Peter leaves Clarissa's house to encounter a representation of space encoding images of Englishness and Empire. Peter has just returned from working as a colonial administrator in India, and as he walks up Whitehall he views a group of boys in uniform, with guns, marching to the Cenotaph.[30] He then walks along Whitehall, replete with statues of military heroes, in a street whose representation of space is one of official governmental power. As the architectural historian Nikolas Pevsner notes, the style of many of these buildings exhibits an 'Edwardian-Imperial optimism'.[31] Woolf represents these streets as markedly masculine in tone, social space devoted to memorials of war, death and Empire, with grand buildings devoted to public, official life, such as the Treasury and the Foreign Office. Halfway along Whitehall, for example, outside the War Office, Peter 'glared at the statue of the Duke of Cambridge' (*MD*, p. 46).[32] This equestrian statue, completed in 1907, is of a former commander-in-chief of the British Army.

Peter's character is thus ambiguously identified with the streets through which he walks. He may glare at Victorian generals, but the image he sees of himself 'in the plate-glass window of a motor-car manufacturer' is that of an administrator of colonial land: 'All India lay behind him; plains, mountains; epidemics of cholera; a district twice as big as Ireland' (*MD*, p. 45). Here Woolf demonstrates the 'ambiguous continuity' between different spaces: the governmental buildings along Whitehall are linked to the India that Peter has administered.[33] Interestingly, India is actually the first place named in the novel, prior to the Westminster of Clarissa's house. The reference to Ireland has a certain topicality, given the setting of *Mrs Dalloway* in June 1923, as the Irish Free State had been established only the year before. If Ireland could leave British control, then the implication was that so might India, a change that Leonard Woolf was actively campaigning for in his anti-imperialist publications in the 1920s.[34] That Peter has put India 'behind' him might be seen as an implied anti-imperialism in the text; but, unfortunately for him, the early part of his walk through the city seems to keep tugging him back into the social spaces of Empire. As Peter 'marched up Whitehall' his gait is echoed in the marching of the boys laying a wreath at the Cenotaph, and the noise of their feet 'drummed his thoughts' (*MD*, p. 46). Peter tries to match their pace, but fails. The boys are treated ambivalently: their faces display 'an expression like the letters of a legend written round the base of a statue praising duty, gratitude, fidelity, love of England' (*MD*, p. 47), transforming their bodies into human statues occupying the Whitehall. The 'life' of the boys, writes Woolf, is smothered beneath 'a pavement of monuments and wreaths and drugged into a stiff yet staring corpse by discipline' (*MD*, p. 47). Yet Peter views them with a faint admiration: 'One had to respect it; one might laugh; but one had to respect it, he thought' (*MD*, p. 47).

Elleke Boehmer, writing of Leonard Woolf's anti-imperialism, notes how in the texts of both Leonard and Virginia Woolf we find 'moments of intense uncertainty' linked to 'colonial disorientation'. For Boehmer such instances pose the question of whether 'modernism cannot be seen as an intrinsic expression of an anxiously imperial world, surveying the breaking apart of trusted cultural certainties'.[35] Peter's response to the marching boys and the Whitehall statues certainly appears as one such moment. A few pages later we detect the same ambivalence, as Peter muses upon coming from

> a respectable Anglo-Indian family which for at least three generations had administered the affairs of a continent (it's strange, he thought, what a sentiment I have about that, disliking India, and empire, and army as he did) there were moments when civilization, even of this sort, seemed dear to him as a personal possession; moments of pride in England. (*MD*, p. 50)

Woolf's use of the brackets seems to be a stylistic device to render the uncertainty Peter feels about his relation to Empire and England. The comment in brackets is a kind of spatial aside, an idea Peter cannot quite fit into the main train of his thought. His 'pride' in this version of Englishness thus textually encloses his reservations about Empire, as if the surrounding social space, of public monuments to one form of Englishness, crowd in upon his personal reticence.

Peter's reaction to the imperial geography of this part of London thus exhibits a confused understanding of his own relation to Empire:

> There they go, thought Peter Walsh, pausing at the edge of the pavement; and all the exalted statues, Nelson, Gordon, Havelock, the black, the spectacular images of great soldiers stood looking ahead of them, as if they too had made the same renunciation (Peter Walsh felt he, too, had made it, the great renunciation), trampled under the same

> temptations, and achieved at length a marble stare. But the stare Peter Walsh did not want for himself in the least; though he could respect it in others. (*MD*, p. 47)

Gordon (in Egypt) and Havelock (in India) were both generals with British Empire connections, and Woolf here suggests how imperialism relied upon the 'disciplining' and 'renunciation' of masculinity in order for it to operate. Peter, too, has denied part of his masculinity (bracketed it off, we might say, as Woolf does here textually) and although he now wants to reject it, he seems unable to escape its strictures. As Peter stands symbolically under the statue of Gordon, he recalls how, as a boy, he had 'worshipped' the general.[36] The materiality of the monuments on Whitehall thus becomes a description of conventional masculinity under Empire, reduced to 'a marble stare'.

As Peter Walsh enters the more open public space of Trafalgar Square his mood alters. He now feels that he has 'escaped' and is 'utterly free' (*MD*, p. 48) from his job as imperial administrator, a role that has just been thrust in his face by the streets through which he has just passed. He also feels free from his divorce in India and now 'inside his brain' feels that he 'stood at the opening of endless avenues' (*MD*, p. 48). Peter's psychic space thus seeps into the external space around him, as Trafalgar Square is an intersection of a number of major London thoroughfares. As an intersection it matches Peter's sense that his previous life, represented in the official spaces of Whitehall, can be left behind, like India, and a new direction taken.

Woolf, however, does not let Peter escape completely. For the next image that catches his eye is that of a young woman: 'But she's extraordinarily attractive, he thought, as, walking across Trafalgar Square in the direction of the Haymarket, came a young woman who ... became the very woman he had always had in mind; young, but stately; merry, but discreet; black, but enchanting (*MD*, p. 48). Woolf here implies a link between the 'marble stare' encountered in Whitehall and the lascivious gaze of Peter at the attractive woman. As he follows her, '[s]traightening himself and stealthily fingering his pocket-knife' (*MD*, p. 48), the violence of the war metamorphoses into a phallic image of predatory male sexuality. As Peter follows the woman along the appropriately named 'Cockspur Street', he fantasises that she is not married, is perhaps a prostitute (not 'respectable'), and that the red carnation she wears is 'making her lips red', an image of female sexuality that causes a 'burning' in his eyes. As Peter pursues her along Piccadilly and Regent Street, he imagines himself to be a 'romantic buccaneer', discarding the official role he played in India.

Eventually, after roaming some distance, the woman disappears into a side-street, and Peter realises his 'fun' is over and that the 'escapade' is 'made up, as one makes up the better part of life, he thought – making oneself up; making her up; creating an exquisite amusement, and something more' (*MD*, p. 50).

Peter's awareness of the constructed nature of his identity echoes Clarissa's earlier revelation that her identity derives from an interaction with the city. It also recalls Bloom's gaze at the woman outside the Grosvenor Hotel in *Ulysses*, and a number of the Imagist poems discussed earlier.[37] Woolf, however, is clear to show how gender has organised this instance of *flânerie*. Peter's 'escapade' is an attempt to shrug off the imperial masculinity represented in the streets of Whitehall, but his escape, suggests Woolf, only draws upon a different aspect of that same version of masculinity. Not the 'marble stare' of war monuments, but the roving eye of the male voyeur. For Peter it may be an 'amusement', but Woolf suggests that this experience of urban space always involves 'something more': not now the marble statues of Whitehall, but the woman as object who helps 'make up' the identity of the male *flâneur*.

Transport technologies

Clarissa Dalloway's declaration 'I love walking in London' (*MD*, p. 7) has ensured that the novel has received a certain amount of attention for its perambulatory qualities.[38] By 1923 the relation of women to city spaces had altered considerably from the turn-of-the-century: the sight of single women walking the streets unaccompanied during the day, such as the woman Peter gazes at, was now accepted. *Mrs Dalloway* is certainly a text that explores these urban pleasures, but the insistent motif of transport in the novel, and elsewhere in Woolf's work, produces a quite distinct set of issues concerning gender, space and literary geography. The significant presence of modern vehicles in Woolf's work is a theme critics such as Rachel Bowlby and Gillian Beer have already opened up for discussion, and one which I want to develop in the rest of this chapter.[39] Transport technologies reveal how Woolf conceptualised the relations between space and place, modernity and modernism, in terms of an exhilarating sense of movement.

An initial survey of modernity and technology in *Mrs Dalloway* reveals numerous instances of transport: there is Elizabeth's bus journey and the freedom it seems to offer her; the 'invisible taxi' that, as John Sutherland notes, conveys Clarissa home after shopping;[40] the grey car, probably that of the Prime Minister, which processes through the crowded streets; and the aeroplane above the city, spraying coffee, toffee or KREEMO. Then there is the motorcar of Dr William Bradshaw, 'low, powerful, grey with plain initials interlocked on the panel' (*MD*, p. 84). This is a symbol of prestige, for as Septimus notes: 'The upkeep of that motor car alone must cost him quite a lot' (*MD*, p. 88); it satirises the grey personality of the car's owner. Ironically, the only vehicle in which Septimus travels is the ambulance that removes his dead body and which Peter Walsh describes as 'one of the triumphs of civilization' (*MD*, p. 134).[41]

Woolf's attitude to these vehicles is complex, in keeping with how technology generally was viewed after the First World War. Alan O'Shea notes how there was an 'immense confidence in technology that prevailed at the beginning of the [twentieth] century', summed up in Peter Walsh's praise for the 'the efficiency, the organization' (*MD*, p. 134), of the ambulance. But, adds O'Shea, this 'fetishism of technology was severely dented by the experience of the First World War, and dystopian representations begin to proliferate from the 1920s'.[42] Prior to the war the Futurist leader Marinetti could cheerfully eulogise: 'A racing car whose hood is adorned with great pipes, like serpents of explosive breath – a roaring car that seems to ride on grapeshot is more beautiful than the *Victory of Samothrace*.'[43] After the war it was precisely this connection of technology with the machinery of war that was seen as problematical. 'The Machine Age' became the term used to describe the modernist works of art that reflected the stage of technological development, specifically painting and sculpture, and the nuances of the term in Europe were almost inevitably tainted by the sense that the First World War was the first technological war, with the massive presence of the tank demonstrating these mournful associations.[44]

Early on in *Mrs Dalloway* war and technology are linked when the backfiring of a motorcar in Bond Street punctuates Clarissa's inner monologue like 'a pistol shot in the streets outside' (*MD*, p. 14). Her inner space (she has been meditating upon which flowers to buy inside a shop) is ruptured by the noises from the outside streets, and Clarissa jumps, startled by the 'violent explosion' (*MD*, p. 14). In the motorcar is a man, unnamed, of political importance, of either royal or governmental status, and Woolf associates the car with the recently finished war. For now she introduces the tormented Septimus Warren Smith, suffering from shell shock, for whom the explosive sound of the car produces an anguished reaction. The car brings the surrounding traffic to a standstill, and onlookers gaze in wonder at it; but for Septimus the 'gradual drawing together of everything to one centre' was 'as if some horror had come almost to the surface and was about to burst into flames' (*MD*, p. 15).

The car resurrects the violence of the war, but its occupation of city space is also a centrifugal display of power provoking Woolf to consider national identity in terms of social space. The car echoes the display of imperial power enacted by the viceregal cavalcade in *Ulysses*. For, as the motorcar parades down Piccadilly with an 'air of inscrutable reserve', the face inside the car is described as 'the majesty of England ... the enduring symbol of the state' (*MD*, p. 16). As it leaves Bond Street the passing car creates a powerful effect upon the inhabitants of the street: 'for in all the hat shops and tailors' shops strangers looked at each other and thought of the dead; of the flag; of Empire. In a public-house a Colonial insulted the House of Windsor, which led to words, broken beer glasses, and a general shindy' (*MD*, p. 18). City space overlaps with national space, and with the colonial space of Empire, with the 'shindy' yet again reinforcing the link to violence. In a revealing spatial image of depth, Woolf then notes how 'the surface agitation of the passing car as it sunk grazed something very profound' (*MD*, p. 18). The horizontal movement of the car turns into a vertical exploration of deeply hidden psychic and national spaces. Again, the most interesting point for a literary geography of modernism is the way that multiple spaces are crossed in a few brief paragraphs – from the psychic space of Septimus's shell shock to the identification of Londoners with Empire and Englishness. One space deliberately not revealed is that inside the motorcar. As Gillian Beer notes, the motorcar is distinguished from the sky-writing aeroplane that appears soon after because, while the car's occupant is hidden from the eyes of the public, symbolising an elite presence, the aeroplane's display is more egalitarian and open to all.[45]

Woolf's text is, therefore, aware both of the threatening potential of the technology of transport, and of how the motorcar can symbolically investigate questions of social and cultural geography. However, Woolf is also willing to embrace the pleasures and possibilities transport offers for exploring modernity. Gillian Beer notes of Woolf's use in *Orlando* of modern objects like the aeroplane: 'Accepting technology into everyday life renews the magical.'[46] This was a point of view expressed in Ford Madox Ford's entreaty to young poets to modernise their geographic vision of the world by taking a trip on the top of a bus to uncharted regions of London.[47] A decade or so later Woolf embraced Ford's advice in her fiction, for there are number of key examples of London seen from an omnibus. As Hermione Lee notes, Woolf's family, when young, did not keep a private carriage, unlike others of their class, and so became accustomed to travelling across London by omnibus. Woolf's mother travelled much by bus, and Virginia once recalled a journey with her sister Vanessa, eating 'bath buns driving down Oxford St. on top of the bus'.[48] A description from her diary in 1918 of a trip during sunset from Oxford Street to Victoria Station shows a similar enthusiasm for this mode of travel, with the city surrounding her as a kind of theatrical pageant.[49]

In Woolf's third novel, *Jacob's Room*, there are a number of descriptions of London that draw upon different experiences of travel across the city.[50] It is the first novel in which Woolf employs a recognisably modernist style, telling the fateful story of Jacob Flanders through multiple perspectives, and in which Woolf begins to put into practice her theories on how modern fiction should be composed. In chapter 5 we find the eponymous hero taking a journey by bus from Oxford Street to St Paul's Cathedral. It is a busy day, and Woolf describes a traffic jam where the vehicles were like 'red and blue beads ... run together on the string' (*JR*, p. 53). Woolf is acutely aware of how transport has drastically altered human relationships in the city:

> The motor omnibuses were locked. Mr. Spalding going to the city looked at Mr. Charles Budgeon bound for Shepherd's Bush. The proximity of the omnibuses gave the outside passengers an opportunity to stare into each other's faces. Yet few took advantage of it. Each had his own business to think of. Each had his past shut in him like the leaves of a book known to him by heart; and his friends could only read the title, James Spalding, or Charles Budgeon, and the passengers going the opposite way could read nothing at all – save 'a man with a red moustache,' 'a young man in grey smoking a pipe.' (*JR*, p. 53)

The spatial proximity of people seated on top of two buses going in different directions only emphasises a profound distance between urban travellers, a point made by the sociologist Simmel.[51] Passengers do not engage with one another, encouraging Woolf to emphasise, in the contrast between the leaves of a book and its title, the sharp divide between inner and outer life that she queried repeatedly in her fiction – and theoretically in her famous essay 'Modern fiction', first published in 1919. In that essay Woolf criticised writers such as Arnold Bennett and H. G. Wells for being 'materialists' who were 'concerned not with the spirit but the body' and with external features of human life like the clothing or the 'the solidity of his [characters'] fabric'.[52] Opposed to this proliferation of surface detail, Woolf wishes to plunge into the hidden psychological lives of her characters, much as one reads the inside of a book rather than merely its cover. This surface–depth model is endemic throughout modernism, but what is of interest here is the way in which Woolf uses modern transport to state her preferred model of literary representation.[53]

Interestingly, however, the movement in *Jacob's Room* from the streets into the inner life is blocked – like the stoppage in the traffic itself.[54] Woolf remains, in this paragraph, at the surface level of description until the buses again begin to move: 'The omnibuses jerked on, and every single person felt relief at being a little nearer to his journey's end, though some cajoled themselves past the immediate engagement by promise of indulgence beyond – steak and kidney pudding, drink, or a game of dominoes in the smoky corner of a city restaurant.' (*JR*, p. 54) This is now not the 'fabric of things' that Woolf complained of in Wells and Bennett,[55] but a peep below into the feelings of the passengers as the buses move away. Woolf's complaint about the materialist Edwardian novelists is often couched in terms of containment within some external space, restricting her belief that modern life can only successfully be represented in terms of movement itself.[56] Thus in 'Modern fiction' Woolf berates Bennett for the way that his characters exist confined within a 'well-built villa in the Five Towns' or a hotel room in Brighton.[57] This means that '[l]ife escapes' and 'has *moved* off, or on, and refuses to be *contained* any longer in such ill-fitting vestments as we provide' (my emphasis).[58] Woolf thus embraces movement as the essence of modern life, and the transits of modern transport become transformed into the metaphors of motion employed in 'Modern fiction':

> Look within ... The mind receives a myriad impressions ... From all sides they come, an incessant shower of innumerable atoms; and as they fall, as they shape themselves into the life of Monday or Tuesday, the accent falls differently from of old ... Is it not the task of the novelist to convey this varying, this unknown and uncircumscribed spirit.[59]

To look within is to try to grasp the swirl of modernity represented so physically by the buses and motorcars of the London streets. The 'shower' of the atoms falling opposes the circumscribing spirit of a Wells or a Bennett. When, in the essay, Woolf praises Joyce for capturing something of this new method of writing fiction, the image is once again of motion: 'Let us record the atoms as they fall upon the mind in the order in which they fall'; Joyce is 'concerned at all costs to reveal the flickerings of that innermost flame which flashes its messages through the brain'.[60] To follow Joyce's method, writes Woolf, takes us from a 'sense of being in a bright yet narrow room, confined and shut in', to a sense of being 'enlarged and set free'.[61] The spaces of modernity must thus be represented in a way that avoids Woolf's aesthetic claustrophobia, taking a cue from the tours of modern transport. 'Oh yes', continues Woolf in the passage quoted earlier from *Jacob's Room*, 'human life is very tolerable on the top of an omnibus in Holborn' (*JR*, p. 54). And if not, it was at the very least a good vantage point for observing the spectacle of the city. Journeying on the tops of buses brought new sensations and pleasures, and an impetus for the fluid style of Woolf's fiction; it is, at any rate, a quite different set of meanings for technology from those represented in the back-firing motorcar in *Mrs Dalloway*.

Woolf, however, found that other forms of public transport reminded her of the constrictive fiction of a Wells or Bennett. In *Jacob's Room* the pleasures of the open-top bus are contrasted with the experience of London workers commuting home by tube. From the top of the omnibus the city, if not one's fellow passengers, can at least be engaged with, but on the Underground the city – and life – escapes:

> Beneath the pavement, sunk in the earth, hollow drains lined with yellow light for ever conveyed them this way and that, and large letters upon enamel plates represented in the underworld the parks, squares, and circuses of the upper. 'Marble Arch – Shepherd's Bush' – to the majority the Arch and the Bush are eternally white letters upon a blue ground. Only at one point – it may be Acton, Holloway, Kensal Rise, Caledonian Road – does the name mean shops where you buy things, and houses. (*JR*, pp. 55–6)

A similarly negative view of the effects of the Underground is found in Woolf's diary for 12 January 1924: 'One of these days I mean to write a story about life turning all the faces in a tube carriage grey, sodden, brave, disillusioned.'[62] This story was never composed, but the dismal experience of the tube did find its way into *The Waves* (1931): 'The descent into the Tube was like death. We were cut up, we were dissevered by all those faces and the hollow wind that seemed to roar down there over desert boulders.'[63] Later in the same novel, after the death of Percival – the central but absent character – the same identification of death with the Underground is made. At the Piccadilly Underground station, Jinny feels that above the ground the 'great avenues of civilization meet' and that she is 'in the heart of life'. But her journey down the escalators is felt as a deathly descent: 'the soundless flight of upright bodies down the moving stairs like the pinioned and terrible descent of some army of the dead downwards and the churning of the great engines remorselessly forwarding us'. This gloomy mood only lifts when she thinks of returning to the surface, with 'the superb omnibuses, red and yellow'.[64] Woolf's dislike of the Underground, and a seeming preference for the omnibus, can be interpreted in terms of her aesthetic dislike of fiction that appears claustrophobic. It can also be seen as one of what James Donald describes as the 'psychic and spatial diseases of modernity', a sense that urban life in the early twentieth century was often experienced in terms of a series of 'spatial phobias' such as claustrophobia, agoraphobia and neurasthenia.[65] While the top of a bus offers a pleasing visual panorama for Woolf, the tube appears a space that is artificial and apart from the stimulations of city life.

One of the most extended accounts of metropolitan travel in Woolf's fiction is Elizabeth's journey in *Mrs Dalloway*. Elizabeth travels eastwards from Victoria Street, up Fleet Street and the Strand, to St Paul's Cathedral, where she feels like 'a pioneer, a stray', for 'no Dalloways came down the Strand daily' (*MD*, p. 122). Elizabeth feels free since she has escaped the clutches of the dreary Miss Kilman and the stuffiness of the Army and Navy Stores. Impetuously, she boards a bus at random, taking a seat upstairs. The buses are colourful entities, like 'garish caravans glistening with red and yellow varnish' (*MD*, p. 120), and a sense of illicit pleasures is attached to the trip.[66] This is emphasised in Elizabeth's fanciful image of the bus as 'a pirate':

> The impetuous creature ... started forward, sprang away; she had to hold the rail to steady herself, for a pirate it was, reckless, unscrupulous, bearing down ruthlessly, circumventing dangerously, boldly snatching a passenger, or ignoring a passenger, squeezing eel-like and arrogant in between, and then rushing insolently all sails spread up Whitehall. (*MD*, p. 120)

Here the swift movement of the prose, with short clauses tucked between commas, captures something of the pirate-bus weaving through the streets. Elizabeth's delight in this voyage extends into a more sensuous realm, where body and technology are united:

> And now it was like riding, to be rushing up Whitehall; and to each movement of the omnibus the beautiful body in the fawn-coloured coat responded freely like a rider, like the figure-head of a ship, for the breeze slightly disarrayed her; the heat gave her cheeks the pallor of white painted wood; and her fine eyes, having no eyes to meet, gazed ahead, blank, bright, with the staring, incredible innocence of sculpture. (*MD*, p. 121)

Elizabeth's pleasure in the trip is a tactile one, differing greatly from the emphasis upon the gaze found in Peter Walsh's urban stroll. Elizabeth stares at no fellow passenger, for she is at the front of the bus, and the experience of the bus journey physically alters her appearance into a kind of mast-head of the 'ship'.

Elizabeth's journey is faintly transgressive, for not only has she ventured into areas unknown to the Dalloways, but she encounters buildings that initiate thoughts of her future. She likes the buildings of the Strand, 'quite different here from Westminster' (*MD*, p. 121), her home, and as she alights from the bus at Chancery Lane she is overwhelmed by the bustle of work all around: 'she would like to have a profession. She would become a doctor, a farmer, possibly go into Parliament if she found it necessary, all because of the Strand' (*MD*, p. 121). Elizabeth thus daydreams of a 'profession for a woman',[67] a desire conflicting with her mother's wishes for her future. The journey into unfamiliar London streets distinguishes her as a 'pioneer', not only geographically but in terms of her gender, dreaming that the independence she has just enjoyed on top of the bus could be translated into her social status. However, as she starts enjoying the noise and 'uproar' of Fleet Street, social convention pulls her back: 'Her mother would not like her to be wandering off alone like this. She turned back down the Strand' (*MD*, p. 123). Her piratical pleasures are over and she boards a bus returning to the more sober environs of Westminster. Woolf links the modernity of the bus journey to a sense of liberty and the advancement of women in society.[68] Sitting on top and at the front of the bus clearly has a symbolic importance here, suggesting a space of advanced thinking above that of the quotidian streets below. The top of the bus has the quality of Foucault's heterotopia, 'a place without a place' (OS, p. 24). Imagining the bus as a pirate ship also confirms Foucault's suggestion that boats are typically heterotopic (OS, p. 27). Elizabeth's identification as a pirate, exploring spaces unknown to her and dreaming of an independent future, illustrates how the heterotopia contests and inverts the real spaces of the London streets through which the bus travels. Elizabeth opens up an 'other space' within London, one that gestures away from the restricted space of her mother's Whitehall geography and her plans for her daughter. Ultimately, the journey by bus offers a heterotopic experience because it is not about some static place, but about relations between different sites; here the spaces of the Strand and beyond versus those of Westminster.

Interestingly, Elizabeth's journey has a counterpoint in the novel when Peter Walsh remembers a similar trip he took with Elizabeth's mother, many years earlier. Peter recalls a Clarissa full of dreams, like her daughter, 'all aquiver in those days and such good company, spotting queer little scenes, names, people from the top of a bus, for they used to explore London and bring back bags of treasures from the Caledonian market' (*MD*, p. 135).[69] Woolf's point is to indicate how Clarissa's present life with her husband, a sober Conservative MP, has suppressed such past explorations. Going to the market, in what was then a much more working-class area of the city, took Clarissa away from the closed social geography she inhabits in the novel, and against which her daughter gently rebels. But, like Elizabeth, Peter Walsh recalls how the bus journey produced an important moment of insight for Clarissa:

> It was unsatisfactory, they agreed, how little one knew people. But she said, sitting on the bus going up Shaftesbury Avenue, she felt herself everywhere; not 'here, here, here'; and she tapped the back of the seat; but everywhere. She waved her hand, going up Shaftesbury Avenue. She was all that. So that to know her, or any one, one must seek out the people who completed them; even the places. (*MD*, p. 135).

Clarissa's identity is thus not fixed, she does not exist deictically 'here', but is only 'completed' by her spatialised relationships with other people and places. The bus journey illustrates a sense of identity in transit, existing 'everywhere' rather than in one settled location. Peter's memory is thus an implicit criticism of the present-day Clarissa, who he perceives as now tied to a view of herself as existing only 'here', in the narrow social geography of Westminster. For him, Clarissa has rejected the heterotopic experience of such travels, unlike her daughter, and has embraced a more permanent sense of place.

In *Mrs Dalloway* Woolf uses the technology of transport in a number of ways to explore the spaces of modernity. After *Mrs Dalloway*, Woolf's perception that the various technologies of transport modify our modern identities is itself transformed by a new personal acquisition, a motorcar.

The hermaphrodite Singer

In July 1927 the lives of Virginia and Leonard Woolf were drastically changed by the purchase of a motorcar. This was enabled by the financial success of *To the Lighthouse*: from 1924 to 1934 their joint annual income rose from £1,047 to £3,615.[70] Leonard notes how the car, a second-hand Singer bought for £275, produced 'a great and immediate effect upon the quality and tempo of our life', and that 'nothing ever changed so profoundly my material existence, the mechanism and range of my every-day life, as the possession of a motor car'.[71] The effect upon Virginia Woolf was equally dramatic, as she testified in her diaries. On 11 July she notes that an 'absorbing subject … has filled our thoughts to the exclusion of Clive & Mary & literature & death & life – motor cars … We talk of nothing but cars.'[72] Frederick Pape, a professional chauffeur who had taught Virginia's sister Vanessa Bell to drive, bought a car for the Woolfs and now gave lessons to both Virginia and Leonard. Throughout the summer and autumn of 1927, and through into their holiday in France by car in 1928, Virginia Woolf's diaries and letters enthuse over the car.[73] This represented something of a *volte-face* for Woolf who, only three years earlier, had complained of the baneful effects of the lowering price of motorcars: 'The cheapening of motor-cars is another step towards the ruin of the country road. It is already almost impossible to take one's pleasure walking'.[74]

The ownership of the car brought with it, in Lefebvre's terms, a 'spatial practice' (*PS*, p. 8) of great freedom, offering an additional pleasure to that of walking. Virginia Woolf wrote: 'This is a great opening up in our lives. One may go to Bodiam, to Arundel, explore the Chichester downs, expand that curious thing, the map of the world in ones mind. It will I think demolish loneliness, & may of course imperil complete privacy.'[75] Woolf's attitude here partly chimes with how social and historical commentators have discussed the 'golden age' of motoring in the 1920s and 1930s in Britain, specifically for the commercial and professional middle classes which formed the group with the highest ownership of cars. The pre-war criticisms of Forster and Masterman, who viewed the car as symptomatic of ostentatious wealth, was now no longer accurate.[76] Woolf herself once held similar opinions. In 1915 she had noted critically how, in the West End of London, she would 'look into the motor cars & see the fat grandees inside, like portly jewels in satin cases'.[77] Motorcar ownership was also higher in the southeast of England than in other areas of the country.[78] The motorcar represented independence and an 'opening up' for the middle classes, a chance to go on 'tours' and day-trips into rural areas, freed from the restrictive timetables of the railways or buses. In 1923 Woolf contemplated exploring the Sussex countryside by taking 'a motor bus ride along the downs … [to] see Steyning, & Arundel'.[79] Numerous publications in this period, such as *Country Life*, carried columns offering advice to urban middle-class drivers wishing to tour the countryside; one 1929 book entitled *Car and Country: Week-End Signposts to the Open Road* offered tips on how to find 'hidden villages' and avoid 'the industrial patch'.[80]

The sites to which Woolf expresses a desire to travel in the car are of some interest. As the historian Sean O'Connell argues, car travel in this period enabled the middle classes to visit places that were distinct from the growing mass-market holiday sites for the working classes, such as the seaside location of Blackpool. It also enabled what O'Connell, following the arguments of John Urry in *The Tourist Gaze*, calls a 'form of commodification, that of selected aspects of "English" heritage and landscape' such as sites of natural beauty or historical interest.[81] The motorcar converted geographical spaces into places imbued with specific social and cultural meanings. Envisaging trips to the Downs and Arundel fits this notion perfectly, as does the reference to a trip to Bodiam, a fourteenth-century moated castle in east Sussex, restored by Lord Curzon and left to the nation.[82] Writing to T. S. Eliot – who now seemed to share an interest in cars according to this letter – Woolf provides a snapshot of the excitements of the day-trip: 'our entire life is spent driving, cleaning, dodging in and out of a shed, measuring miles on maps, planning expeditions, going on expeditions, being beaten back by the rain, eating sandwiches on high roads, cursing cows, sheep, bicyclists, and when we are at rest talking of nothing but cars and petrol'.[83]

The motorcar resonates with a set of personal meanings combining social and geographical freedom with the psychic space that was so significant for Woolf. In the revealing phrase Woolf uses in her diary – 'the map of the world in ones mind' – we see the coalescence of different forms of space: psychic, cartographic and geographic. The car signified individual freedom for Woolf, a feeling shared by other people of her class in the 1920s, but it also captured her personal sense of the pleasures and perils of moving through modernity, and has a noteworthy presence in her writing in this period.

Woolf closely associated the dark-blue Singer with her writing, for as she notes: 'The world gave me this for writing The Lighthouse.' It is a 'nice light shut up car in which we can travel thousands of miles',[84] and certainly over the coming weeks, when the Woolfs moved down, for the summer, from London to Rodmell in Sussex, the car assumed a central role in their lives. Early in August she notes: 'We have motored most days. We opened one little window when we bought the gramophone; now another opens with the motor.'[85] The technology of the car had, by the late 1920s, clearly shrugged off the aura of violence associated with the war. Woolf's entry for 10 August continues to eulogise the emancipations of the car:

> Yes, the motor is turning out the joy of our lives, an additional life, free & mobile & airy to live alongside our usual stationary industry. We spin off to Falmer, ride over the Downs, drop into Rottingdean, then sweep over to Seaford, call, in pouring rain at Charleston ... return for tea, all as light & easy as a hawk in air. Soon we shall look back at our pre-motor days as we do now at our days in the caves.[86]

Here Woolf tacitly links the mobility of the car to the modernity of the times. In a contemporaneous letter to Lytton Strachey, she amusingly suggested that the horrors of the Victorian period were perhaps 'explicable by the fact that they walked, or sat behind stout sweating horses'.[87] The technology of the car is a form of 'industry' that marks modern times, and shows how the very fact of movement itself, as well as the places one visits, became a key pleasure for Woolf. If *To the Lighthouse* was a novel based around the completion of one personal journey, the Singer car, product of that book's success, brings the promise of multiple excursions. Not only does the car modify the quotidian life of Woolf but, in offering an 'additional life', it emphasised one of Woolf's key themes, that of the narrowness of a unitary self. The mobility of the car also began to infuse the 'stationary industry' of Woolf's writing, demonstrating an impact upon the style of her fiction.

Touring by car engaged a different geography, mainly of the country rather than the city found in *Mrs Dalloway*. In August 1927 Woolf witnessed a 'scene' while on a motor journey in Sussex that she wanted to record:

> One was on the flats towards Ripe one blazing hot day. We stopped in a bye road about 3 in the afternoon, & heard hymn singing. It was very lonely & desolate. Here were people singing to themselves, in the hot afternoon. I looked & saw a middle class 'lady' in skirt & coat & ribboned hat, by the cottage door. She was making the daughters of the agricultural labourers sing; it was about three o'clock on a Tuesday perhaps. Later we passed the ladies [*sic*] house; it had a wooden griffin nailed above the door – presumably her crest.[88]

Woolf links this incongruous incident to the spatial practice of motoring itself, and concludes by noting the need for a deeper historical understanding of the practice:

> What I like, or one of the things I like, about motoring is the sense it gives one of lighting accidentally, like a voyager who touches another planet with the tip of his toe, upon scenes which would have gone on, have always gone on, will go on, unrecorded, save for this chance glimpse. Then it seems to me I am allowed to see the heart of the world uncovered for a moment. It strikes me that the hymn singing in the flats went on precisely so in Cromwell's time.[89]

Woolf's reflection here is a complex one upon history and the process of recording 'moments' from life pregnant with significance, both personally and socially. Woolf's fiction is, of course, full of this desire to arrest the instant, the 'moment of being', a kind of modernist sublime in which the particular takes on a more historical or transcendent resonance. What is of interest here is the way Woolf connects this moment to the spatial explorations of the motorcar. It recalls Woolf's description of her fictional technique as a 'tunnelling process, by which I tell the past by instalments'.[90] Motoring is a form of overground exploration, visiting sites whose historical geography can be uncovered by the viewer. The car symbolises an absolutely modern experience, conveyed in the image of the voyager to another planet, with its connotations of a future time contained in the now; but the car simultaneously also allows a glimpse into a past world, of a cultural practice Woolf believes can be traced back over the centuries. It is a particularly *English* spatial history that is being considered here, implied also in the picture of Woolf as explorer, the metropolitan intellectual in anthropological mode, stumbling upon an 'unrecorded' incident from English common life, unchanged since the seventeenth century. These thoughts rely upon the practice of motoring to sites, and demonstrate how Woolf perceives certain places to contain a sedimented spatial history that might be excavated further.

Woolf's pleasure in the motorcar had a lengthy genealogy in her personal history. The diary entry that follows the description of hymn singing at Ripe finds Woolf remembering how, a year earlier, in 1926, she had been very depressed, her state coinciding with the completion of *To the Lighthouse*. On 28 September she noted an 'intense depression' that she had felt periodically for some time, and gave an account of an argument with Leonard about work on the garden at Monks House, in Sussex. Woolf records saying to Leonard that if they hired a full-time gardener, 'we shall be tying ourselves to come here; shall never travel; & it will be assumed that Monks House is the hub of the world. This it certainly is not, I said, to me.'[91] Woolf's desire for travel and to avoid being tied to one place is linked, in her mind, to the depression she has been suffering. To evade this depression in the future she prescribes herself a series of measures: 'first, incessant brain activity; reading, & planning; second, a methodical system of inviting people here ... third, increased mobility. For next year, I shall arrange perhaps to go definitely to Ethel Sands. With my motor I shall be more mobile.'[92] Ethel Sands, hostess and patron of the arts, lived in a lesbian household in Dieppe, and Woolf did visit her the following summer, partly alleviating her craving for travel. But the motorcar represents a deeper yearning than that for a mere vehicle enabling day-trips and holidays. Movement is the key to staving off depression and a sense of restriction, with the motorcar being a physical manifestation of 'incessant brain activity'. In 1924 she thought she might cure a minor depression by 'crossing the channel' to Dieppe, or 'by exploring Sussex on a motor bus'.[93] Later, in 1933, the Woolfs purchased a new car, a Lanchester, that Virginia nicknamed 'The Deluge', and she described how her mind began to work like the engine of the car: 'a Rolls Royce engine ... powering its 70 miles an hour in my brain'.[94] Woolf's awareness that she needed 'increased mobility' when unhappy was also a reaction to her unpleasant memories of the 'rest cures' she undertook in 1910 and 1912, when confinement was prescribed and mental exertion was prohibited.[95] The motorcar represents another escape from fixed categories for Woolf, operating as a kind of heterotopia to subvert a too narrow sense of place. Woolf preferred a more spatial understanding of her environment, tours rather than maps in de Certeau's terms.

By this time Woolf had commenced her relationship with Vita Sackville-West, a 'flamboyant' driver with whom she often travelled by car and who had given Woolf an early driving lesson.[96] Soon after their first meeting Vita visited Woolf in her 'large new blue Austin car, which she manages consummately'.[97] Female motorists had increased in number a little throughout the 1920s, although they were often decried for being interested in cars only as fashionable objects of consumption. In 1927 the magazine *Motor Trader* remarked upon a 'fashion for driving cars' that was developing among women, only to compare it with women's interest in 'a new fashion in hats or frocks'.[98] On 20 September 1927, Woolf first conceived the idea for *Orlando* (published 1928), with Vita as the model for the hero(ine). Woolf told Vita that the description of her driving in *Orlando* showed 'the most profound and secret side of your character'.[99] Woolf linked the androgyny explored in *Orlando* with her Singer motorcar in letters to her sister Vanessa: 'I can't believe your amazing stories of the Male and Female parts of the Renault. Do the French sexualise their engines? The Singer I know for a fact to be hermaphrodite, like the poet Cowper.'[100]

In *Orlando* Vita is memorably portrayed as an 'expert driver' (*O*, p. 200).[101] At the wheel of a car in central London she drives through the streets with road rage *avant la lettre* in the 'present day' of 1928:

> [P]eople crossed without looking where they were going. People buzzed and hummed round the plate-glass windows within which one could see a glow of red, a blaze of yellow, as if they were bees, Orlando thought – but her thought that they were bees was violently snipped off and she saw, regaining perspective with one flick of her eye, that they were bodies. 'Why don't you look where you're going?' she snapped out. (*O*, pp. 195–6)

Orlando's impatience with the bodies in the streets not only reveals something of her character, but indicates how the human body was already spatially subservient to the motorcar in European cities by the late 1920s. The architect Le Corbusier, for example, in his influential modernist manifesto *The City of Tomorrow* (1925), wrote of Paris: 'To leave your house meant that once you had crossed your threshold you were a possible sacrifice to death in the shape of innumerable motors.'[102] Orlando's savage 'flick of her eye' emphasises how motoring had also altered the visual apprehension of fellow inhabitants in the city, here being drained of all colour. After shopping at a department store Orlando drives off, with another salvo of abuse for pedestrians and fellow drivers: 'she frowned slightly, changed her gears admirably, and cried out, as before, "Look where you're going!" "Don't you know your own mind?" "Why didn't you say so then?" while the motor car shot, swung, squeezed, and slid ... down Regent Street, down Haymarket, down Northumberland Avenue, over Westminster Bridge' (*O*, p. 200).

Orlando's trip takes her over the river and along the Old Kent Road, into the poorer southeast of London, a region very little represented in Woolf's fiction. Woolf has a keen eye for the social geography of this part of London. In *A Room of One's Own* Woolf used this area of London to imagine a historiography that traced the 'infinitely obscure lives' of women; and in the same book the fictional Shakespeare's sister is said to be buried 'at some cross-roads where the omnibuses now stop outside the Elephant and Castle'.[103]

In this section of *Orlando* Woolf's writing eschews a more fluid style for a jagged, plainer prose designed to capture the piecemeal nature of the motorist's perceptions of the surrounding scenes: 'People spilt off the pavement. There were women with shopping bags. Children ran out. There were sales at drapers' shops. The streets widened and narrowed. Long vistas steadily shrunk together. Here was a market. Here a funeral' (*O*, p. 200). Woolf's point, however, seems to be a more general one about the nature of the motorist's relations to the space of the city. Orlando sees fragments of text, adverts for shops: 'Applejohn and Applebed, Undert-. Nothing could be seen whole or read from start to finish. What was seen begun – like two friends starting to meet each other across the street – was never seen ended' (*O*, p. 200). This description is very much in agreement with Simmel's observations on the disjointed visual relationships that dominate in cities, where people move mainly in machines and not of their own volition.[104] The scrambled words also recall the fragmented signification of the aeroplane sky-writing in *Mrs Dalloway*. In *Orlando* Woolf reflects upon how motoring is at the root of this process of heterotopic disintegration, particularly directed at subjectivity. Now it is not the texts of shop signs that are fragmented; rather it is the human body and mind that become like 'scraps of torn paper tumbling from a sack'. The implication, Woolf writes, is that 'the process of motoring fast out of London so much resembles the chopping up small of identity which precedes unconsciousness and perhaps death that it is an open question in what sense Orlando can be said to have existed at the present moment' (*O*, pp. 200–1). Orlando reassembles herself only when she leaves the city and starts to see 'green screens' of countryside against which 'the little bits of paper fell more slowly', which gradually lead once more to the illusion of unity. This image of a divided self echoes the historically distinct selves of Orlando's lives over the centuries of the book, but it is significant that the motion of the car produces this particular spatial dissolution of identity. It is a further image of the 'flux' of modernity diagnosed by Forster nearly twenty years earlier. Woolf's image, however, though it is compared to the moment prior to unconsciousness and maybe death, seems somehow more benign and less fearful than Forster's in *Howards End*.

As Orlando continues her journey towards her ancestral home in Kent (based on Knole, Vita Sackville West's home) and the conclusion of the book, the motorcar precipitates a swift dash through the multiple selves she has possessed: her identity is in transit, 'changing her selves as quickly as she drove' (*O*, p. 202). New selves appear at every corner that the car passes, a fluctuation of identity that Woolf argues occurs when 'the conscious self, which is the uppermost, and has the power to desire, wishes to be nothing but one self' (*O*, p. 202). Woolf's true self is a 'compact of all the selves we have it in us to be', and this myriad identity is 'commanded and locked up by the Captain self, the Key self, which amalgamates and controls them all' (*O*, p. 202). The car journey has unleashed these multiples selves, spinning them out from the controlling Captain, as if propelled by the wheels of the car. Once again the flux of modernity is activated by the motorcar, here as an agent that enters into the castle of identity only to dissolve it. The self becomes like snapshots of the surrounding landscape taken from the speeding car, and almost impossible to synthesise into a more unified picture.

Woolf then provides a lengthy paragraph illustrating this process, where Orlando drives to seek her true self and the narrator offers a modest commentary: 'we only copy her words as she spoke them, adding in brackets which self in our opinion is speaking, but in this we may well be wrong' (*O*, p. 203). The effect of this technique is to create a very dramatic form of textual space:

> Trees, she said. (Here another self came.) I love trees (she was passing a clump) growing there a thousand years. And barns (she passed a tumble-down barn at the edge of the road). And sheep dogs (here one came trotting across the road. She carefully avoided it). People? (She repeated it as a question.) I don't know. Chattering, spiteful, always telling lies. (Here she turned into the High Street of her native town, which was crowded, for it was market day, with farmers, and shepherd, and old women with hens in baskets.) I like peasants. I understand crops. But (here another self came skipping over the top of her mind like the beam from a lighthouse). Fame! (She laughed.) Fame! Seven editions. A prize. Photographs in the evening papers (here she alluded to the 'Oak Tree' and 'The Burdett Coutts' Memorial Prize which she had won; and we must snatch space to remark how discomposing it is for her biographer that this culmination to which the whole book moved, this peroration with which the book was to end, should be dashed from us on a laugh casually like this; but the truth is that when we write of a woman, everything is out of place – culminations and perorations; the accent never falls where it does with a man). (*O*, pp. 203–4)

This paragraph is interesting for several reasons. First, because a number of the comments in brackets refer not to a particular self but describe a place or scene, and seem to suggest that the boundaries between psychic identity and physical place are here being elided. Second, because the asides create a quite distinctive spatial style. It appears as an authorial commentary on the character's thoughts, forming a kind of parallel location for the narrator herself, turning the textual space of the novel into another vertiginous experience. The words in brackets somehow exist embedded in the landscape through which Orlando travels, popping up when her gaze alights upon them, or flashing back from the landscape to the viewer, as the lighthouse image here suggests. The multiple selves of Orlando's history are transformed into the many spaces the text explores: the interior space of Orlando's thoughts, the places she/he looks at, and the bracketed spaces of the author's commentary. Between these diverse spaces a number of dialogic relations, in Bakhtin's sense, are produced: as the part of the final sentence in brackets implies, the narrator seems to jostle for space with her creation.[105] The perpetual movement of Bakthinian dialogism seems very apt here, paralleling the motion of Orlando's motorcar. Woolf's final witty lament, that her biographical subject has escaped her, reveals an interesting gendering of space and place. Orlando flees from the author's culminating words, for a woman is always 'out of place': femininity exists heterotopically, akin to the travelling car and the kaleidoscopic selves, rather than as a form of identity located by a static discourse of place.

The association of manifold subjectivities with motoring is also found in a short and cryptic essay composed by Woolf around the time, in 1927, when she and Leonard first enjoyed driving around Sussex. Like *Orlando*, 'Evenings over Sussex: reflections in a motor car' investigates the notion of multiple selves. The essay starts as a kind of topographic account of the Sussex landscape and its towns, but quickly widens its focus to the spatial history of the region. Sussex is playfully imagined as an old woman, glad when night can shade her face like a veil, only to reveal a more flattering outline. Woolf comments on how the sea cliffs still remain, even after the sights to be found at various towns are hidden by the 'veil of evening': 'All Eastbourne, all Bexhill, all St Leonards, their parades and their placards and their invalids and char-à-bancs, are all obliterated.'[106] Woolf, so often noted for her descriptions of city space, here produces an interesting set of pastoral images:

> What remains is what there was when William came over from France ten centuries ago: a line of cliffs running out to sea. Also the fields are redeemed. The freckle of red villas on the coast is washed over by a thin lucid lake of brown air, in which they and their redness are drowned. It was still too early for lamps; and too early for stars.[107]

Signs of contemporary civilisation are removed as Woolf pictures the landscape as it was, emphasising how all space, as Lefebvre argues, is the product of human activity. As with the incident of people singing near Ripe, Woolf reads the landscape in terms of its spatial history.

The next paragraphs shift perspective from the object viewed to the subject who perceives. Interestingly, the viewer is not situated in one spot, as in many traditional topographic forms of writing. Woolf thus eschews a discourse of place, in de Certeau's terms, for an actualisation of space as movement, for this is an essay that reports a tour through space, rather than producing a mapped discourse of the region. Looking at the landscape also undermines the integrity of the gazing subject: 'one looks up, one is overcome by beauty extravagantly greater than one could expect … and then, when all seems blown to its fullest and tautest, with beauty and beauty and beauty, a pin pricks; it collapses'.[108] Woolf appears dissatisfied that she cannot represent this beautiful scene, complaining: 'I cannot hold this – I cannot express this – I am overcome by it – I am mastered.'[109] As Rachel Bowlby notes, this inability is linked to the romantic experience of the sublime, in which the writer must attempt to represent the unrepresentable, in order to confirm the subjectivity of the viewer rather than be 'mastered' by the landscape.[110] As Woolf notes, her 'discontent' was linked to 'the idea that one's nature demands mastery over all that it receives; and mastery here meant the power to convey what one saw now over Sussex so that another person could share it'.[111]

Woolf's strategy in responding to this dilemma is to reassemble the subjectivity of the viewer into several selves. As in *Orlando*, travelling through the countryside in the motorcar occasions this proliferation of selves. Woolf notes how 'another prick of the pin' occurs when one becomes aware of the rapidity of the passing scenes: 'one was wasting one's chances; for beauty spread at one's right hand, at one's left; at one's back too; it was escaping all the time; one could only offer a thimble to a torrent that could fill baths, lakes'.[112] Woolf thus decides to introject this flow of external scenery by forming a 'torrent' of new identities: 'relinquish, I said (it is well known how in circumstances like these the self splits up and one self is eager and dissatisfied and the other stern and philosophical), relinquish these impossible aspirations; be content with the view in front of us'.[113]

These two selves now conduct a dialogue with one another about how to address the beauty they perceive 'as the car sped along'. Woolf now creates a third persona, an 'I' that is 'aloof and melancholy' who converts the experience of travel through space into a journey through time: 'I feel life left behind even as the road is left behind. We have been over that stretch, and are already forgotten.' This melancholy lament is finally interrupted by a fourth self that draws attention not to the passing of time, but to the future in the image of a flickering light in the landscape: 'You, erratic and impulsive self that you are, feel that the light over the downs there emerging dangles from the future ... I feel suddenly attached not to the past but to the future. I think of Sussex in five hundred years to come.'[114] This light of the future conjures up a technological realm of 'magic gates' and '[d]raughts fan-blown by electric power' that clean houses (vacuum cleaners, perhaps). Central to this technologised future is the electric light of the motorcar: 'Look at the moving light in that hill; it is the headlight of a car. By day and by night Sussex in five centuries will be full of charming thoughts, quick, effective beams.'[115] The inner space of 'charming thoughts' thus merges with the external space illuminated by the beams of the car's headlights.

Woolf's narrator now tries to sum up this kaleidoscope of ideas as the light fades and the landscape dims from view. It is time 'to collect ourselves; we have got to be one self.' Woolf's narrator acts as a central ego, like the Key Self in *Orlando*, reviewing the images and impressions the car journey has provided and attempting to make sense of them:

> Let me see; there was a great deal of beauty brought in today; farmhouses; cliffs standing out to sea; marbled fields; mottled fields; red feathered skies; all that. Also there was disappearance and the death of the individual. The vanishing road and the window lit for a second and then dark. And then there was the sudden dancing light, that was hung in the future. What we have made then today ... is this: that beauty; death of the individual; and the future.[116]

The narrator's picture of the day is parodied as a kind of account keeping (of the sort that Leonard Woolf was obsessed by),[117] and is then turned into a solid form, 'a little figure for your satisfaction', a kind of doll sitting on the knee of the narrator. It is a puzzling image, suggesting that Woolf herself is unsure of what the day's perceptions amount to, that the desire to represent landscape is bound, as in the sublime, to escape, even with the help of multiple selves. Beauty, death and modernity coalesce here, with the 'little figure' gesturing towards something profound, 'as if the reality of things were displayed there'. The viewing selves suddenly experience another moment of sublimity: 'A violent thrill ran through us; as if a charge of electricity had entered into us. We cried out together: "Yes, yes," as if affirming something, in a moment of recognition.'[118] The electricity of modernity itself – of the car headlights – seems to flood through the multiple selves, but the insight cannot be put further into words, much like Lily Briscoe's final 'vision' in *To the Lighthouse*.

The essay concludes with the appearance of a different self, that of the body, which speaks with a noise 'as low as the rush of the wheels' of the car. The body wishes to return home to eat and go to bed, and so the narrator dismisses the 'assembled selves' to enjoy the rest of the journey 'in the delicious society of my own body'.[119] 'Evenings over Sussex' is a mysterious yet inconclusive piece of writing, a kind of self-reflective musing by a novelist on the processes surrounding the creation of beauty. What is striking, however, is the fashion in which it is structured around a dialogue between flux and form, from the fluid traffic of different selves, across scenes of Sussex and the motions of the car, to the final static form of the 'little figure' of the doll, a kind of mannequin of modernity representing what the novelist might craft with the fluctuating images just explored. It is an attempt to circumvent the failure of the romantic sublime by replacing it with a modernist 'representational space' as fluidity captured in form: place as a stable locus is thus overwhelmed by the spatial explorations of the motorcar.

Leonard Woolf's comment on the effects of foreign travel on Virginia perhaps explains why the essay on the Sussex countryside is so tentative:

> She had a passion for travelling, and travel had a curious and deep effect upon her. When she was abroad, she fell into a strange state of passive alertness. She allowed all these foreign sounds and sights to stream through her mind; I used to say rather like a whale lets the sea-water stream through its mouth, straining from it for its use the edible flora and fauna of the seas. Virginia strained off and stored in her mind those sounds and sights, echoes and visions, which months afterwards would become food for her imagination and her art. This and the mere mechanism and kaleidoscope of travel gave her intense pleasure, a mixture of exhilaration and relaxation.[120]

The essay is thus an example of this meditative stage of creation, of the 'kaleidoscope of travel' prior to final conceptualisation. The significant point is how the motorcar acts as the carrier for a variety of experiences of modernity. From the 1920s onwards Woolf became more aware of the cultural ramifications of transport as a key bearer of modernity, and sought to integrate the spatial experiences of travel by bus and motorcar into her writings. This is also seen in her major essay 'Modern fiction'. When Woolf searches in the material world for an analogy to the creation of modern literature, in the 1925 version she substitutes the motorcar for the bicycle of the 1919 version.[121] Not surprisingly, this essay argues that modern life cannot be re-imagined in terms of such an antiquated form of transport as a horse-drawn carriage with its 'gig lamps'.

Another significant incident demonstrating Woolf's use of motor transport occurs in *A Room of One's Own*, when the narrator looks from her London window and sees a man and a woman get into a taxi.[122] The cab then glides off 'as if swept on by the current'[123] of urban life. This is the trigger for Woolf's famous and controversial articulation of androgyny, where the sexual division of mental life is unified: 'when I saw the couple get into the taxi-cab the mind felt as if, after being divided, it had come together again in a natural fusion'.[124] As Woolf notes, the 'sight was ordinary enough', but what felt odd was 'the rhythmical order with which my imagination had invested it'.[125] Like her hermaphrodite Singer, Woolf seems to invest the taxi with the capacity to unsettle established categories of thought, making it too a heterotopia. Urban transport suggests the fluidity of the androgynous mind, with the car that shifts through space unfixing the sexual identities housed in the brain. The internal space of the mind and the external space of the city streets once again interact, rhythmically, to produce Woolf's theory of androgyny. The taxi is an image of spatial and social trespass, representing a challenge to the sense of power over space typified at the start of the book, when the female narrator trespasses on the grass of an Oxbridge college only to be upbraided by a male beadle.

Woolf's version of literary geography is not an inert mapping of place; like Joyce's, it is a series of tours that records the spatial history of locations and draws upon the social and political geography of particular places. Equally, it shifts between 'houses in the brain' and physical locations such as the city or the countryside. Within Woolf's geographic imagination the figure of transport produces a kaleidoscopic sense of the modern self that she embraced for its potential to unsettle fixed structures of power. The car and the bus as quotidian technologies of modernity by the 1920s can be seen to influence her style, much as did the cinema. However, the difference between these technologies is that the motorcar represents a much more haptic experience of modernity, shown in Elizabeth's bus-top journey, Orlando's furious drive through the city, or the role of the body in 'Evening over Sussex'. With Woolf, then, we discover a modernism committed to exploring and expanding what she called 'that curious thing, the map of the world in ones mind'.

In the next chapter we discover another writer, Jean Rhys, who is committed to exploring a psychic 'map of the world', but who does so in a fashion much more problematic than that found in Woolf.

Notes

1 These books were part of a larger practice of literary geography in this period, a project that aimed to chart London as a literary and historical location, for the heritage interests of the London County Council. For an excellent overview of this form of literary geography see Andrea P. Zemgulys, '"*Night and Day* is dead": Virginia Woolf in London "Literary and Historic"', *Twentieth Century Literature* 46: 1 (spring 2000), 56–77.

2 Virginia Woolf, 'Literary geography', *The Essays of Virginia Woolf*, 6 vols, ed. Andrew McNeillie (London, Hogarth Press, 1986), vol. 1: *1904–1912*, p. 33.

3 Woolf, 'Literary geography', p. 35.

4 I have concentrated on this period of Woolf's work in order to continue a rough chronology for examining modernism's engagement with space and geography, and to provide a manageable amount of material to discuss. Clearly, texts like *The Years* (1937) and *Between the Acts* (1941) could be analysed in terms of their exploration of a form of spatial history.

5 Virginia Woolf, *Jacob's Room*, ed. Sue Roe (London, Penguin, 1992).

6 Woolf, *The Diary of Virginia Woolf*, 5 vols, ed. Anne Oliver Bell and Andrew McNeillie (London, Penguin, 1982); vol. 3, November 1928, p. 209.

7 Randall Stevenson, *Modernist Fiction: An Introduction* (Hemel Hempstead, Harvester Wheatsheaf, 1992), p. 58. For an account which focuses upon the

formal aspect of Woolf spatial vision, particularly the connections with the visual arts, see Jack F. Stewart, 'Spatial form and color in *The Waves*', *Twentieth Century Literature* 28: 1 (spring 1982), 86–107.

8 Gillian Beer, *Virginia Woolf: The Common Ground* (Edinburgh, Edinburgh University Press, 1996), pp. 52–3.

9 Woolf, 'Chateau and country life' (1908), *Essays*, vol. 1, p. 222.

10 'The mark on the wall' (1917), in *A Haunted House and Other Stories* (Harmondsworth, Penguin, 1973), pp. 44–5.

11 For Donald, in *Mrs Dalloway* the city is seen 'through a cinematic structure of visibility'; see James Donald, 'This, here, now: imagining the modern city', in Sallie Westwood and John Williams, eds, *Imaging Cities: Scripts, Signs, Memories* (London and New York, Routledge, 1997), p. 188. For essays on other aspects of Woolf's engagement with technology see Pamela L. Caughie, ed., *Virginia Woolf in the Age of Mechanical Reproduction* (New York and London, Garland, 2000).

12 Brenda R. Silver, ed., *Virginia Woolf's Reading Notebooks* (Princeton, NJ, Princeton University Press, 1983), p. 240. I am grateful to the University of Sussex Special Collections for allowing me to view a copy of this map. Silver argues that Woolf initially composed the map for the short story 'Mrs. Dalloway in Bond Street', published in 1923 in the *Dial*. In another piece of autobiographical writing, 'Old Bloomsbury', Woolf recalled first moving to the Bloomsbury area of London: her sister Vanessa was 'looking at a map of London and seeing how far apart they were (–) had decided that we should leave Kensington and start life afresh in Bloomsbury'; see Virginia Woolf, *Moments of Being: Unpublished Autobiographical Writings*, ed. Jeanne Schulkind (Sussex, University Press, 1976), p. 162.

13 Woolf, 'Portraits of places' (1906), *Essays*, vol. 1, pp. 124–5.

14 See, for example, the map of 'Central London in the mid-twenties' in the 1992 Penguin editions of *Jacob's Room* and *Mrs Dalloway*; the OUP edition of *Ulysses* also follows this helpful editorial practice with its reproduction of a contemporary map of Dublin.

15 See chapter 1 for this distinction by de Certeau.

16 Virginia Woolf, *The Letters of Virginia Woolf*, 6 vols, ed. Nigel Nicholson and Joanne Trautmann (London, Hogarth Press, 1975–80), vol. 6, p. 434.

17 Susan M. Squier, 'Virginia Woolf's London and the feminist revision of modernism', in Mary Ann Caws, ed., *City Images: Perspectives from Literature, Philosophy, and Film* (New York, Gordon & Breach, 1991), p. 99. For detailed treatments of this theme see Susan M. Squier, *Virginia Woolf and London: The Sexual Politics of the City* (Chapel Hill and London, University of North Carolina Press, 1985) and Dorothy Brewster, *Virginia Woolf's London* (New York, New York University Press, 1960). Woolf herself wrote a number of essays on the topography of London in the late 1920s and early 1930s, the most important of which are 'Street haunting: a London adventure' and the

series of essays with the title 'The London scene': see *The Crowded Dance of Modern Life: Selected Essays, volume 2*, ed. Rachel Bowlby (London, Penguin, 1993).

18 The phrase 'intellectual aristocracy' comes from Noel Annan, *Our Age: Portrait of a Generation* (London, Weidenfeld & Nicholson, 1990). For a critical discussion of Bloomsbury as a 'fraction' of an enlightened social class see Raymond Williams, 'The Bloomsbury fraction' in *Problems in Materialism and Culture* (London, Verso, 1980).

19 Virginia Woolf, *Mrs Dalloway* (London, Granada, 1976).

20 Woolf, *Diary*, vol. 2, 3 November 1923, p. 272.

21 Another interpretation might be that it reflects Woolf's problem with putting other spaces into novelistic representation. We can, perhaps discount this if we consider the topographical sketches in her series of essays of 1931–32 'The London scene' (in Woolf, *The Crowded Dance of Modern Life*), which contains a much more nuanced urban geography.

22 Wyndham Lewis, 'Manifesto', *BLAST* 1 (1914), ed. Bradford Morrow, reprinted (Santa Rosa, CA, Black Sparrow Press, 1989), p. 18.

23 Woolf, *Diary*, vol. 2, p. 15.

24 Woolf, 'Street haunting', *Crowded Dance of Modern Life*, pp. 78–9. Sometimes the northern suburbs of London are similarly depicted: perhaps the most damning put-down of Charles Tansley in *To the Lighthouse* (1927) (London, Granada, 1977), p. 181, is the statement: 'He had married; he lived at Golder's Green.'

25 Woolf, *Diary*, vol. 1, p. 9.

26 Woolf, *Diary*, vol. 1, pp. 29–30.

27 Woolf, *Diary*, vol. 2, p. 301.

28 Woolf, *Diary*, vol. 2, p. 298.

29 Elizabeth Wilson, *The Sphinx in the City: Urban Life, the Control of Disorder and Women* (London, Virago, 1991), pp. 23–4.

30 The Cenotaph, memorial to the First World War dead, had only recently been completed, in 1920, designed by the leading 'imperial' British architect, Sir Edwin Lutyens.

31 Nikolas Pevsner, *The Buildings of England: London I – The Cities of London and Westminster* (Harmondsworth, Penguin, 1957), p. 471.

32 Woolf also features this statue in *A Room of One's Own* (1929) (London, Granada, 1977) as an example of how the 'trophies and cannon' (p. 38) of male power are celebrated in public spaces in London.

33 The government buildings at the bottom of Whitehall used to house the India Office.

34 In this period Leonard Woolf published *Economic Imperialism* (1920) and *Imperialism and Civilisation* (1928), as well as being involved in the establishment of the League of Nations. For further details see Patrick Brantlinger, '"The Bloomsbury Fraction" versus war and empire', in Carola M. Kaplan and

Anne B. Simpson, eds, *Seeing Double: Revisioning Edwardian and Modernist Literature* (Basingstoke, Macmillan, 1996).

35 Elleke Boehmer, '"Immeasurable strangeness" in imperial times: Leonard Woolf and W. B. Yeats', in Howard J. Booth and Nigel Rigby, eds, *Modernism and Empire* (Manchester and New York, Manchester University Press, 2000).

36 Woolf sometimes shared the puzzling emotions Peter experiences here. In March 1926 she went to Greenwich and 'almost burst into tears over the coat Nelson wore at Trafalgar': *Diary*, vol. 3, p. 72. When her novel *Night and Day* was published (21 October 1919) she noted that it was also Trafalgar Day, the anniversary of Nelson's death: *Diary*, vol. 1, p. 306.

37 Later in the novel Peter perceives London itself as being like 'a woman who had slipped off her print dress and white apron to array herself in blue and pearls' (p. 143).

38 See, for example, Rachel Bowlby's 'Walking, women and writing' in her *Feminist Destinations and Further Essays on Virginia Woolf* (Edinburgh, Edinburgh University Press, 1997). For Woolf's own *flâneur* text, with its account of 'the greatest pleasure of town life in winter – rambling the streets of London' (p. 70), see 'Street haunting', in *Crowded Dance of Modern Life*. For an excellent discussion of this essay and *Mrs Dalloway* in terms of the relationship between the city and modernity see Laura Marcus, *Virginia Woolf* (Plymouth, Northcote House, 1997), ch. 4.

39 See ch. 1 of Bowlby, *Feminist Destinations*, for a brilliant reading of Woolf's essay 'Mr Bennett and Mrs Brown' that stresses the significance of its setting on a train from Richmond to Waterloo. Gillian Beer provides an excellent history of the symbol of the aeroplane in Woolf's writing in 'The island and the aeroplane', in Homi K. Bhabha, ed., *Nation and Narration* (London and New York, Routledge, 1990).

40 John Sutherland, 'Clarissa's invisible taxi', in *Can Jane Eyre Be Happy?* (Oxford, Oxford University Press, 1997).

41 The ambulance, from the French *hôpital ambulant* (walking hospital), does seem a recent invention. The *Oxford English Dictionary* lists the first usage as 1809, referring to a moving hospital during war, and notes that the term only became common during the Crimean War of 1854–56. The use of ambulances for civilian wounded is first cited only in 1922.

42 Alan O'Shea, 'English subjects of modernity', in Mica Nava and Alan O'Shea eds, *Modern Times: Reflections on a Century of English Modernity* (London, Routledge, 1996), p. 18.

43 F. T. Marinetti, 'The founding and manifesto of Futurism' (1909), in *Futurist Manifestos*, ed. Umbro Apollonio (London, Thames & Hudson, 1973), p. 21.

44 For the role of the tank in modernism see Trudi Tate, *Modernism, History and the First World War* (Manchester, Manchester University Press, 1998), ch. 5. In a number of places Woolf describes the city itself as a machine, but often with connotations of activity or industry. For example, she writes in

A Room of One's Own that 'London was like a workshop, London was like a machine' (p. 27).

45 As Beer also notes, the aeroplane in *Mrs Dalloway* is – in contrast to the motor-car – not linked to the war; see Gillian Beer, 'The island and the aeroplane', pp. 274–6.

46 Beer, 'The island and aeroplane', p. 278.

47 Ford Madox Hueffer (Ford), 'Modern poetry', *The Thrush* 1: 1 (December 1909), 51, 53.

48 Hermione Lee, *Virginia Woolf* (London: Vintage, 1997), p. 37.

49 Woolf, *Diary*, vol. 1, p. 111.

50 Marcus notes how the view from the top of a bus operates as a 'counterpoint to the sensations and impressions of the walker in the city': Marcus, *Virginia Woolf*, p. 62.

51 See the brief discussion of Simmel's views on transport in chapter 3.

52 Woolf, 'Modern fiction' (1925) in *The Crowded Dance of Modern Life*, pp. 6–7.

53 For a consideration of the philosophical model of surface versus depth within other modernists see Sanford Schwartz, *The Matrix Of Modernism: Pound, Eliot, and Early Twentieth Century Thought* (Princeton, NJ, Princeton University Press, 1985).

54 Traffic is again blocked by two significant processions at the end of *Jacob's Room* – one pro- and one anti-war (pp. 150–3).

55 The phrase 'the fabric of things' occurs in the companion essay to 'Modern fiction' entitled 'Mr Bennett and Mrs Brown', in Woolf, *A Woman's Essays: Selected Essays, volume 1*, ed. Rachel Bowlby (London, Penguine, 1992), p. 82.

56 A point also noted by Bowlby, *Feminist Destinations*, p. 5.

57 Woolf, 'Modern fiction', pp. 6–7.

58 Woolf, 'Modern fiction', p. 7.

59 Woolf, 'Modern fiction', p. 8.

60 Woolf, 'Modern fiction', p. 9.

61 Woolf, 'Modern fiction', pp. 9–10.

62 Woolf, *Diary*, vol. 2, p. 286.

63 Virginia Woolf, *The Waves* (London, Granada, 1977), p. 120. For a discussion of how the idea of being underground has been imagined metaphorically in literature see Rosalind Williams, *Notes on the Underground. An Essay on Technology, Society, and the Imagination* (Cambridge, MA, MIT Press, 1990).

64 Woolf, *The Waves*, p. 131.

65 James Donald, 'Imagining the modern city', pp. 193–4.

66 A possible precursor for Elizabeth's bus journey might be the account in Dorothy Richardson's *Backwater* (1916) volume of *Pilgrimage*. Here Miriam Henderson, also on top and at the front of a bus, travels into central London with her mother; see Dorothy M. Richardson, *Pilgrimage* (London, Virago, 1979), vol. 1, pp. 193–9.

67 The title of a Woolf lecture from 1931.

68 A point also made by Marcus, *Virginia Woolf*, p. 78.

69 The Caledonian Market in Islington was the site of an ancient cattle market; it thrived in the early years of the twentieth century, and the painter Walter Sickert, about whom Woolf wrote an essay, once declared that the Caledonian Market was his idea of heaven; quoted in Ann Saunders, *The Art and Architecture of London: An Illustrated Guide* (Oxford, Phaidon, 1984), p. 297.

70 See Leonard Woolf, *Downhill All the Way: An Autobiography of the Years 1919–1939* (London, Hogarth Press, 1967), p. 177. I only discovered two excellent essays discussing Woolf and the motorcar after completing the manuscript of this book: see Makiko Minow-Pinkney, 'Flanerie by motor car?', in Laura Davies and Jeanette McVicker, eds, *Virginia Woolf and Her Influences: Selected Papers from the 7th Annual Conference on Virginia Woolf* (New York, Pace University Press, 1998), and the same author's 'Virginia Woolf and the age of motor cars', in Pamela L. Caughie, ed., *Virginia Woolf in the Age of Mechanical Reproduction* (New York and London, Garland, 2000).

71 Leonard Woolf, *Downhill All the Way*, pp. 177–8.

72 Woolf, *Diary*, vol. 3, p. 146.

73 Only Leonard seems to have persevered, passing his driving test early in 1928. Woolf does talk of driving from 'the Embankment to the Marble Arch and only knocked one boy very gently off his bicycle. But I would rather have a gift for motoring than anything': Letter to Ethel Sands, 22 July 1927, in *Letters*, vol. 3, p. 400. Two days later she confesses to 'wobbling round and round Windmill Hill, every day, trying to avoid dogs and children': Letter to Janet Case, 24 July 1927, in *Letters*, vol. 3, p. 403.

74 Woolf, 'The cheapening of motor-cars' (1924) in *The Essays of Virginia Woolf*, vol. 3 , p. 440.

75 Woolf, *Diary*, vol. 3, p. 147.

76 See chapter 2, this book, for this criticism.

77 Woolf, *Diary*, vol. 1, p. 17.

78 Sean O'Connell, *The Car and British Society: Class, Gender and Motoring, 1896–1939* (Manchester and New York, Manchester University Press, 1998), p. 85.

79 Woolf, *Diary*, vol. 2, p. 259.

80 See O'Connell, *The Car and British Society*, pp. 154–5; and see ch. 5 for O'Connell's thorough consideration of the impact of the car in rural areas.

81 O'Connell, *The Car and British Society*, p. 79. See John Urry, *The Tourist Gaze: Leisure and Travel in Contemporary Societies* (London, Sage, 1990).

82 Pevsner calls Bodiam 'the ideal picture of the powerful ... planned castle'; see Ian Nairn and Nikolaus Pevsner, *The Buildings of England: Sussex* (Harmondsworth, Penguin, 1965), p. 419.

83 Letter to T. S. Eliot, 24 August 1927, in *Letters*, vol. 3, p. 413.

84 Woolf, *Diary*, vol. 3, p. 147. This model was later replaced by a coffee-and-chocolate-coloured Sun Singer in February 1929 (with sun-roof), and then by a larger silver-green Lanchester 18 in 1933.
85 Woolf, *Diary*, vol. 3, p. 151.
86 Woolf, *Diary*, vol. 3, p. 151.
87 Woolf, *Letters*, vol. 3, 3 September 1927, p. 418.
88 Woolf, *Diary*, vol. 3, p. 153.
89 Woolf, *Diary*, vol. 3, p. 153.
90 Woolf, *Diary*, vol. 2, p. 272.
91 Woolf, *Diary*, vol. 3, p. 112.
92 Woolf, *Diary*, vol. 3, p. 112.
93 Woolf, *Diary*, vol. 2, p. 308.
94 Woolf, *Diary*, vol. 4, p. 142. This marked the beginning of the composition of Woolf's novel *The Years*.
95 For information on Woolf's 'rest cures' see Lee, *Virginia Woolf*, ch. 10. For further information on the social history of the 'rest cure' see Elaine Showalter, *The Female Malady: Women, Madness and English Culture, 1830–1980* (London, Virago, 1987).
96 The description of Sackville-West's driving comes from Leonard Woolf, *Downhill All the Way*, p. 112. For the story of the driving lesson in Richmond Park see Lee, *Virginia Woolf*, p. 509. The most thorough account of Woolf's relationship with Vita Sackville-West is found in Suzanne Raitt, *Vita and Virginia: The Work and Friendship of Vita Sackville-West and Virginia Woolf* (Oxford, Clarendon Press, 1993).
97 Woolf, *Diary*, vol. 2, p. 313.
98 Quoted in O'Connell, *The Car in British Society*, p. 67.
99 Woolf to Vita Sackville-West, 6 March 1928, in *Letters*, vol. 3, p. 469.
100 Woolf to Vanessa Bell, 21 Feb. 1928, in *Letters*, vol. 3, p. 463. The analogy with Cowper remained in Woolf's mind for, in *A Room of One's Own* (1929), he is one of the select group of male writers who is said to possess an androgynous mind (p. 98). It is Cowper's poem 'The Castaway' that Mr Ramsey repeatedly utters in *To the Lighthouse*.
101 Woolf, *Orlando* (London, Vintage, 1992).
102 Le Corbusier, *The City of Tomorrow and its Planning*, trans. F. Etchells (1925) (New York, Dover, 1987), p. xxiii. Le Corbusier's memorable response to this threat was to argue for the exclusion of the pedestrian from city-centre roads.
103 Woolf, *Room of One's Own*, pp. 85 and 47.
104 See Georg Simmel, 'Soziologie des Raumes', quoted in Walter Benjamin, *Charles Baudelaire: A Lyric Poet in the Era of High Capitalism*, trans. Harry Zohn (London, Verso, 1983), pp. 37–8.
105 Bakhtin's conception of dialogic relations is found in his 'Discourse in the novel' in Mikhail Bakhtin, *The Dialogic Imagination: Four Essays*, ed. Michael

Holquist, trans. Caryl Emerson and Michael Holquist (Austin, University of Texas Press, 1981).

106 Woolf, 'Evening over Sussex: reflections in a motor car', in *Crowded Dance of Modern Life*, p. 82.

107 Woolf, 'Evening over Sussex', p. 82. Lee notes that Woolf produced some of 'the most pastoral city novels ever written': *Virginia Woolf*, p. 421. *Mrs Dalloway*, for example, though often seen as predominantly urban is, in fact, split between London and the many scenes from Clarissa's memory, set at Bourton, in the countryside.

108 Woolf, 'Evening over Sussex', p. 82.

109 'Evening over Sussex', p. 82.

110 See Bowlby, Notes to *Crowded Dance of Modern Life*, p. 199.

111 'Evening over Sussex', p. 82.

112 'Evening over Sussex', pp. 82–3.

113 'Evening over Sussex', p. 83.

114 'Evening over Sussex', p. 83.

115 'Evening over Sussex', p. 84.

116 'Evening over Sussex', p. 84.

117 See Lee, *Virginia Woolf*, ch. 17, for this aspect of Leonard's character.

118 'Evening over Sussex', p. 84.

119 'Evening over Sussex', p. 85.

120 Leonard Woolf, *Downhill All the Way*, pp. 178–9; and for an account of their first trip by car, to Cassis in March 1928, see pp. 181–5. An eclectic collection of Woolf's writings is found in *Travels with Virginia Woolf*, ed. Jan Morris (London, Pimlico, 1997). The book is a somewhat of a curio, as Morris notes, since '[n]obody was ever less of a travel writer, in the usual sense of the phrase, than Virginia Woolf' (p. 1). However, Morris argues: 'Few writers have ever been more powerfully inspired by the sense of place' (p. 3). This, as I have argued, is a sense of place that opens up the social and historical resonances of the geographies represented in her writings.

121 Woolf, 'Modern fiction' in, *Crowded Dance of Modern Life*, p. 5.

122 For a careful examination of the ambiguous implications of this passage see Bowlby, *Feminist Destinations*, pp. 35–9.

123 Woolf, *A Room*, p. 92.

124 Woolf, *A Room*, p. 93.

125 Woolf, *A Room*, p. 92.

6 The voyages of Jean Rhys

In *Wide Sargasso Sea* (1966), Jean Rhys's most acclaimed novel, the heroine Antoinette Cosway is brought by boat to England from the Caribbean, but feels that somehow during the voyage 'we changed course and lost our way to England'.[1] This perception of a voyage that goes adrift from its final destination might be taken as a dominant trope for all of Rhys's work. In her fiction of the 1930s we find a very different dialectic of space and place from that articulated in the writing of modernists in the previous two decades. By discussing one novel in detail, *Voyage in the Dark* (1934), we can see how Rhys's work exhibits a passage through modernity that constantly subverts any discourse of place as settled attachment.[2] In Rhys the quest for the fixity of place is always undermined by a spatial history determined by two key features: her experience of being a woman alone in the cities of London and Paris;[3] and her status as a colonial exile from Dominica, a place whose imperial history can be traced throughout her texts. For Rhys the voyage of subjectivity is always a little off course, and never arrives at its destination.

Whereas Woolf and Joyce explored the geography of London and Dublin in order to demonstrate how any particular location can reveal a complex spatial history, Rhys's work reverses this focus, as she starts from a bewildering experience of spatial flux and seeks to return to a place of fixity that is forever absent. This perpetual shuttling between spaces is noticeable in the way that Rhys's texts make much use of liminal spaces, such as cafés or restaurants. Her heroines appear to hover between inner and outer spaces, in a state of geographic ambivalence. The wider spatial form of many of her novels also demonstrates this feature: *Voyage in the Dark* exists between the geographies of London and Dominica, whereas *After Leaving Mr Mackenzie* (1930) is split between Paris and London. *Wide Sargasso Sea* replicates the distance between the West Indies and England in the different narrative points of view employed in the novel. Textually, then, these novels represent the motif of the voyage; the voyage *out*, as in the passage from England to the colony explored in the story of Rochester in *Wide Sargasso Sea*; or the voyage *in*, such as Anna's travel to the imperial centre in *Voyage in the Dark* or Antoinette's in *Wide Sargasso*

Sea. Crucially, however, there is also the voyage *between*, noticeably in Julia's movement between Paris and London in *After Leaving Mr Mackenzie* and Anna's psychic travels between England and Dominica.

The image of a journey between different locations is perhaps the dominant one in Rhys's fiction, shown in the manner in which Julia and Anna shift from one grubby lodging-house to the next. To move between locations implies that one never quite settles – one is always between spaces. The typical Rhys heroine never really occupies anywhere, never 'dwells', in Heidegger's sense: a hotel room is only ever a kind of temporary halt.[4] In *Quartet* (1928) Marya Zelli reflects that her life in Paris lacks 'solidity; it lacked the necessary background. A bedroom, balcony and *cabinet de toilette* in a cheap Montmartre hotel cannot possibly be called a solid background.'[5] Like Benjamin's *flâneur*, for whom the city street or café becomes a form of interior, Rhys's characters feel most at home in places of fleeting location on the borders between inside and outside spaces.[6] On the first page of *Quartet*, for example, we learn that Marya has just spent nearly an hour-and-a-half in a Parisian café. Another example of this focus upon borderline spaces is found in *After Leaving Mr Mackenzie*, where one short chapter takes place on a staircase inside Julia's lodgings. This existence in liminal spaces, however, only parallels the broader location of Rhys's fiction between the geographies of the Caribbean and Europe. Much Rhys criticism has tended to focus upon one of two geographies in her work: *either* the metropolitan centres of Paris, London and Vienna, *or* the Caribbean influences upon her fiction.[7] This chapter argues that a fuller understanding of Rhys's modernism must examine how her texts voyage between these two geographies, capturing the shifting shapes of her own prose journeys.

The 1930s are often viewed as the golden age of literary travel writing, and it is illuminating to compare Rhys's voyages with this non-fictional genre. Many writers in the 1930s endeavoured to escape the feeling that the earth was now a thoroughly known place. D. H. Lawrence, for example, complained in 1931: 'Superficially, the world has become small and known … We've done the globe and the globe is done.'[8] Lévi-Strauss echoed this complaint when commenting on the travels undertaken in the 1930s that became the source of his *Tristes Tropiques*: 'I wished I had lived in the days of real journeys, when it was still possible to see the full splendour of a spectacle that had not yet been blighted, polluted and spoilt.'[9] Writing of interwar British literary travel writing, Paul Fussell distinguishes between explorer, traveller and tourist by suggesting that the explorer 'seeks the undiscovered' and 'moves towards the risks of the formless and the unknown', whereas the tourist seeks that which has been discovered by business and 'moves towards the security of pure cliché'. The traveller exists between these two poles, 'retaining all he can of the excitement of the unpredictable attaching to exploration, and fusing that with the pleasure of "knowing where one is" belonging to tourism'.[10] In that sense writers such as Graham Greene, in *Journey Without Maps* (1936), sought to represent their travels as if they were still explorers, moving through uncharted territory, rather than through a familiar and known place.[11] Rhys, however, does not fit the categories established by Fussell, who considers only male writers. She is neither a tourist nor a traveller, although her fiction does tend towards 'the formless and unknown' mode of the explorer, stumbling through a landscape without an adequate map. As Rachel Bowlby notes of Sasha Jensen in *Good Morning, Midnight* (1939), Rhys's protagonists are types of 'negative *flâneuse*', who tread the streets to fill in empty time rather than to take pleasurable strolls through London as exhibited in Woolf's *Mrs Dalloway*.[12]

Rhys's fiction thus offers an alternative literary geography to that of travel writing in the 1930s. It is fictional, though often thinly disguised, autobiography, and is not primarily intended to depict place in the mode of travel writing.[13] However, her texts do travel through spaces, depicting how her heroines experience particular places. As Rhys noted in her autobiography, 'The place I live in is terribly important to me, it always has been.'[14] Her novels are, thus, very precise in their naming of particular boarding-houses, cafés, restaurants and streets in London and Paris. *After Leaving Mr Mackenzie*, for example, has five chapters named after specific areas of London and Paris. Deborah Parsons offers a careful and illuminating account of how Rhys's texts revise the image of the city found in earlier twentieth-century urban fiction by utilising the 'negative *flâneuse*' figure noted by Bowlby. For Parsons the foremost image in Rhys's work is that of being 'lost in a labyrinth', a woman struggling to survive in the city who often wishes to retreat from the open streets.[15] Particular urban places are also important to Rhys's characters; but, as Parsons suggests, they 'have no claim on these places for identity … the places themselves are paradoxically places of non-place, places of the dispossessed'.[16] Rhys's heroines are thrown into the heterotopic flux of the city, travelling between spaces, and never able to convert these spaces into places of belonging. This, we might say, is a form of anti-travel writing, where the travel is enforced, always between sites and with no return home.

Rhys's rather negative representational spaces of the city need to be connected to the imperial spaces of her Dominican home, and to the formal strategies of her writing, in order to produce a more comprehensive understanding of her literary geography. In one short section of

Culture and Imperialism entitled 'A note on modernism', Edward Said links the formal innovations of literary modernism with the historically changing geography of European imperialism in the late nineteenth and early twentieth centuries. This is a very suggestive hermeneutical route into Rhys's fiction. Said argues that an 'unsettling anxiety'[17] began to beset European metropolitan culture in the early years of the century, revolving around the notion that imperial empires were no longer invulnerable monoliths but had to deal with – what Said terms – 'the contending native and the fact of other empires', factors that might undermine the dominant visions of imperialism. European culture faced these tensions and threats ironically, suggests Said, 'with a desperate attempt at a new inclusiveness',[18] and it is that ambition which underpins a number of key modernist innovations. Said notes that modernism's central response to empires took the shape of 'a new encyclopaedic form' with three distinctive features: a use of circular structures that were both open and closed simultaneously, evident in such works as *Ulysses* and *The Cantos*; a rendering of the experience of newness that deliberately drew upon fragments and references to previous cultures, shown in Joyce's adaptation of the *Odyssey*; and a cultural form 'that draws attention to itself as substituting art ... for the once-possible synthesis of the world empires'.[19] It is this final characteristic that seems most applicable to the modernism of someone like Rhys, a writer who uses only the first two characteristics, and then in a more circumscribed fashion.[20] Said's elaboration of this idea of the substitution of art for empire is worth quoting:

> When you can no longer assume that Britannia will rule the waves forever, you have to reconceive reality as something that can be held together by you the artist, in history rather than in geography. Spatiality becomes, ironically, the characteristic of an aesthetic rather than of political domination, as more and more regions – from India to Africa to the Caribbean – challenge the classical empires and their cultures.[21]

Said's rather compressed point is that the spatial practices of imperialism are projected into the aesthetics of modernism because of an awareness that the actual material spaces of empires are threatened. Said's argument draws, on one level, on Joseph Frank's reading of modernist form, namely that modernist writers such as Joyce and Eliot employ a 'spatial form' in their texts as a new way of treating temporality in narrative.[22] Said, however, wishes to extend Frank's conception of spatial form into the realm of 'social space', connecting the formal spaces of the literary text to actual material spaces such as buildings, cities and countries. In Said's view modernism's metaphorical exploration of literary space is a substitution for the imperial exploration of material spaces and territories. This point indeed illuminates Rhys's fiction, where the investigation of a whole variety of spaces – psychic, metropolitan, colonial – is of fundamental importance. However, Said's argument can be refined by showing that Rhys's texts do not simply replace metaphorical for material space, but rather constantly shift between the two, turning geographical locations into landscapes charged with symbolic value, and treating metaphorical spaces as if they really existed. Utilising de Certeau's distinction between space and place, as in previous chapters, generates a more nuanced account of how Rhys's work combines spatial forms with material geographies.

The textual space found in *Voyage in the Dark* consists of three aspects: the narrative structure of the novel, with its movement between two geographical locations; the peculiar attention given to certain representational spaces such as cafés, restaurants or hotel rooms; and the spatial style of Rhys's prose, with its many short sentences, short paragraphs and multiple sections to each part or chapter producing an effect of multiple fragments surrounded by empty spaces. This chapter revolves around these three aspects, beginning with a consideration of the presence of the imperial geography of Dominica.

Mapping the Caribs and the 'Imperial Road'

Rhys was born to a British colonial family in 1890 in Dominica, an island in the West Indies colonised first by the French and then in 1805 by the British, and notable for having one of the few sizeable populations of native American Indians, the Caribs, left in the West Indian islands. Rhys's family was part of the very small white presence left on the island, and she was sent to England in 1907 at the age of 16. Rhys returned to Dominica only once, in 1936. That holiday visit came just two years after the publication of *Voyage in the Dark*, the novel which, aside from the later *Wide Sargasso Sea*, is concerned most with her home and with post-colonial notions of belonging and exile. In the novel we can see how the experiences of home, homecoming and displacement are mapped on to her depiction of the spaces of London, and how those experiences illuminate Rhys's particular engagement with modernity.

If Rhys's return voyage to 'the only home I ever had'[23] was prompted partly by the experience of writing a novel about a heroine, Anna Morgan, who has left Dominica for a miserable life in London, it is of interest that Rhys's holiday also included a visit to unfamiliar terrain. For it was the only time she visited the Carib quarter of the island. It is as if writing a novel about failing to belong in a particular place produced a desire to visit a location and a people who have perhaps the strongest claim to being the autochthonous inhabitants of Dominica. The Caribs of Dominica exist as a kind of shadowy presence in Rhys's fiction, serving as a reminder of how important particular geographies are for the psychic notion of belonging. The Carib quarter also indicates how European imperialism, as has often been noted, relied upon the conquest of particular spaces. The Caribs, who have resisted complete colonisation by Europeans since being first encountered by Columbus in 1492, testify to how skirmishes over certain geographies continue to inform the backdrop of much twentieth-century writing.

Having said something about the spatial history of the Caribs, I turn to the short story 'Temps perdi', started during Rhys's trip to Dominica. I wish to demonstrate how Rhys's literary voyages between different spaces are a complex response both to the imperial geography of her native island, and to her encounter with the metropolitan centres of London and Paris. In this way we can locate Rhys within a new spatial history of modernist writing. This is a spatial history, in Paul Carter's sense, that 'discovers and explores the lacuna left by imperial history [and] begins and ends in language'.[24] In other words, the experimental form of Rhys's fiction must be linked to the spatial history of her construction as a writer, a quite different history quite different from that, say, of Woolf in London.

The very existence of the Carib quarter illustrates one of the spatial lacunae deriving from the imperial history of the island. The Carib Reserve (now Carib Territory) was the creation of the Governor General of Dominica Hesketh-Bell, a friend of the Rhys family.[25] Towards the end of the nineteenth century Dominica was taken up by the British Premier Joseph Chamberlain, as an example of the modernising of the administration of British colonies. This was part of the 'new imperialism' that then dominated European politics. Hesketh-Bell was sent on an official visit to the Carib quarter in 1902 to ascertain the utility of officially recognising the area as distinct from the rest of the island. The Caribs had occupied only a small section of the north-east of the island since 1763 when Britain gained control of the island from the French. Also in 1763 the island was surveyed and divided into lots for auction in London, with one small area being preserved for the Caribs. Hesketh-Bell noted that white estate owners often tried to extend the borders of their properties into the Carib Territory, but met resistance from the Caribs. We see here a struggle over territorial occupation, but also a marginalising of the spatial presence of the Caribs. Hesketh-Bell sought a final mapping of the scope of the Caribs' lands, partly because of an ethnological interest in what was perceived as a dying 'race', and possibly influenced by the typical Victorian fear of miscegenation, but also for reasons of more efficient imperial administration. Hesketh-Bell noted of the Caribs that it 'is to be regretted, from an ethnological point of view, that the breed is suffering much from the admixture of negro blood'.[26] He also argued that it 'appeared to me very desirable that the limits of the Reserve should be properly and finally delimited' and 'highly desirable that the small remnant of the people who once owned the whole island ... should be permanently guaranteed the possession of their last homes'.[27] Hesketh-Bell thus commissioned a new survey of Carib land and suggested a small extension to the territory assigned to them. Hesketh-Bell's plan to map and control space typifies the imperialist project of delimiting borders – paralleling the dislike of 'racial' miscegenation – and is a good example of Lefebvre's sense of an official 'representation of space'. It also reminds us how Rhys's writing transgresses these kinds of boundaries, moving across 'racialised' and national spaces. As Maggie Humm notes in relation to Rhys's fiction, an 'insistent theme in colonial writing is its tendency to abstract relationships as mappable geographic space', for 'the map is the colonial signifier of a dominated race, its economy, and topography'.[28]

The link between cartography and imperialism at the turn of the nineteenth century has been well explored.[29] Said, for example, notes: 'Imperialism and the culture associated with it affirm both the primacy of geography and an ideology about control of territory.'[30] And as Derek Gregory notes, not only was cartography crucial to imperialist expansion, but geography as a discipline 'was prompted to some extent by the imperatives of colonialism and imperialism'.[31] At the inaugural meeting of the Scottish Geographical Society in December 1884, for example, H. M. Stanley praised the way in which the discipline 'has been and is intimately connected with the growth of the British Empire'.[32] One key example in the British context was the work of Halford Mackinder, a keen supporter of Chamberlain's campaign for imperial protection against free trade. One of Mackinder's most influential works was his textbook *Britain and the British Seas* (1902) in which he argued that one important consequence of the 'new world' discoveries of Columbus was a re-orientation of Britain's location in the geographical world-order: 'Britain became the central, rather than the terminal, land of the world.'[33] Mackinder's book positions Britain at the geopolitical centre of the world, a relocation he uses to strongly justify the British Empire: 'Hence the most important facts of contemporary political geography are the extent of the red patches of British dominion upon the map of the world.'[34] Interestingly, Mackinder notes the intrinsic connection between imperial domains abroad and the metropolitan character of London. Not only would 'Metropolitan England' (Mackinder's term for the London metropolis) be 'poorer but for foreign investments and imperial rule' but 'much of its governmental and financial activity, and no small part of its residential character' can be traced to 'the imperial rank of London'.[35]

It is this spatial history of British imperialism that forms the background to much of Rhys's work, offering a way of understanding her particular voyages through a modernity constructed between the metropolis and the colony. This is shown in Rhys's short story 'Temps Perdi', which fictionalises her 1936 visit to Salybia in the Carib quarter. The story is told by means of two flashbacks, one to Vienna and one to Dominica, by an unnamed female narrator who longs to escape from her confining existence in a small English village. The means of escape is memory, a device that is also prominent throughout *Voyage in the Dark*. Remembered places are more vivid and appealing than the present and provide a way for the heroine to try to break away from what is imaged as her repetitive life where 'the days ... wait around the corner to be lived again'.[36] This image of time's circularity (the title translates as the Creole for 'wasted time', in addition to a sense of 'lost time') is reinforced in a passage leading up to the visit to the Carib quarter. The narrator starts to reminisce about a dress she once had:

> And thinking of it I am free again, knowing that nobody can stop me thinking, thinking of my dresses, of mirrors and pictures, of stones and clouds and mountains and the days that wait for you round the corner to be lived again. Riding round and round the Inner Circle, but unlike Matsu I ride knowing that it will be dark and cold when I come out, that it will be November, and that I shall be a savage person – a real Carib.[37]

This complex fantasy is interesting for the connection Rhys makes between the enclosed and circular metropolitan space of the London Underground railway (the Inner Circle) and the space of her memories of the Dominican Caribs. That this is a fantasy of flight is made apparent in the next lines when the narrator remembers another facet of the Caribs: 'They run and hide when they see anybody ... Perhaps I shall do that too.'[38] The image of the Underground (see chapters 3 and 5, this book) is a key feature of modernist London, but here is a metaphor for being trapped in a certain place: the Caribs symbolise an escape from the enclosed space, shown in their capacity to hide from view. This ability is demonstrated during the visit the story describes, since the two Caribs, a mother and daughter, met by the narrator during that visit seem not to be genuine Caribs, lacking their supposed physical features. The actual geography of the Carib quarter aids this representation of flight, for the reserve exists in a most inaccessible part of the island. To Rhys the Carib's existence represents the possibility of a group that has broken out of the 'wasted time' of modern life, and found a different ordering of space. The Carib reserve, then, functions as a heterotopic space of escape for Rhys, a Foucauldian counter-site that draws its strength from the geographical and historical marginality of the actual quarter on the island.[39] A link to Said's argument about modernism and imperialism is also established in this association of a specific social space with a narrative problematising temporality by means of its circular structure, and showing Rhys's typical theme of the repetitiveness of life.[40]

Memory in this story, and elsewhere in her writings, is a peculiar kind of space for Rhys, and the voyage into memory is often an escape from the problematical exterior world. Remembered places seem more material than the present, making the psychic space of memory a location to which Rhys obsessively returns. Often it appears that Rhys's heroines are engaged in a drive to incorporate outside spaces, trying to bring inside that which exists in outside reality, as in the notion of the 'savage' and secret Carib Territory in 'Temps Perdi'. Another example of this incorporation of external space is found in the story of Rhys's search, while on holiday in 1936, for the so-called 'Imperial Road' that crossed Dominica from the west to the east coast. Rhys insisted to her taxi driver that the road still existed, even though he said that it had become overgrown and disused. Rhys persisted in trying to discover the road and, even after she and her husband completed the journey – by mule along a steep mountain track – refused to accept that the road no longer existed, choosing to believe instead that she had been lied to.[41] Rhys refused to accept exterior reality and could not adjust to this geographical amendment to her mental landscape; instead, her psychic space had to incorporate and refigure the external world. It is another example of the many 'spatial phobias of modernity' noted by James Donald in relation to Woolf's *Mrs Dalloway*; the difference here is that the principal context is not the metropolis but an imperial domain.[42] Rhys's strange psychosis also recalls one component of the geography of imperialism itself: to remap conquered territories as part of the process of asserting the power of the new owner over the landscape.[43]

Dominica's Imperial Road was constructed at the end of the nineteenth century by Hesketh-Bell as part of the economic expansion planned by the 'new imperialism'. In Rhys's story 'Pioneers, Oh, pioneers', the stated aim of the Imperial Road was 'to attract young Englishmen with capital who would buy and develop properties in the interior' of the island.[44] Teresa O'Connor, in a discussion of Rhys's unpublished short story 'The Imperial Road' has suggested that the road was never actually finished, due to exorbitant costs and inclement conditions. The mythical

Imperial Road, O'Connor argues, had two functions for Rhys: first, it signified an emblem of home and childhood, of a concrete place that could anchor her desire to belong; and, second, it was 'a metaphor for colonialism, its disappearance signifying the end and failure of colonialism'.[45] For the narrator of the story, and for Rhys also, this disappearance represents a loss of identity. The non-existence of the road results in a dangerous disturbance in the complex cartography of Rhys's psyche: the metaphorical space of the remembered Imperial Road must be prioritised over the material space of the absent road. Here we see another reminder of Said's argument about modernism and imperialism, with the modernist writer substituting the exploration of imagined spaces for the unproblematical imperial occupation of the actual lands and territories. Rhys's longer fiction, however, is more nuanced than this argument appears to suggest. Her treatment of different spaces does not rely simply on the replacement of material for imagined space, or of memory for real places; rather, a novel like *Voyage in the Dark* exhibits a bewildering movement across and between these different spaces and locations, inhabiting the flux of modernity diagnosed in *Howards End* and which I have traced in Joyce, Woolf and the Imagists. It is this disorientation that connects Rhys's representations of Dominica's colonial history with a central feature of the spatial history of modernism not fully discussed by Said: the experience of the city.

This is England

Commenting on the intricate relations between modernism and Empire, Patrick Williams wonders whether writers such as Rhys and Katherine Mansfield should be regarded as creating a distinct modernism of Empire, or whether they introduce an imperial history into the construction of metropolitan modernism.[46] The ambiguous trajectory of Rhys's modernism is particularly apparent when we examine how urban space is represented in *Voyage in the Dark*. One initial difficulty with Rhys's portrayal of London in that novel lies in deciding whether the alienated and displaced life of the modernist city as experienced by her young heroine, the white West Indian Anna Morgan, is just a new form of Rhys's primal feeling of not belonging, or whether the city experience of dislocation throws into relief the feeling of exile. If the latter view dominates, then the city encounter with alienation, exile and polyglossia is actually what presents Rhys with a voice and a vocabulary for these sensations of estrangement.[47] It is difficult to decide which of these two readings of Rhys and the city dominates in the novel, and it is perhaps better to understand them dialogically, as another form of narrative movement in the text. In this sense the city is a metaphor for other experiences of being a stranger, as well as a concrete presence in her fiction, offering locations from which Rhys can begin to ponder the idea of 'home' and place. The key to resolving this paradoxical representation of the city lies in tracing the way that Rhys's texts constantly traverse limits and borders, exemplified in the way that Anna Morgan's consciousness in *Voyage in the Dark* oscillates between her present life in London and her past experiences in Dominica.

Unable to locate herself either in Dominica or England, Rhys's uncertaintly about her identity seems to be mapped on to her depiction of particular urban spaces.[48] But this makes it sound too neat, as if psychic unease about one's identity is simply figured in a set of external tropes. Instead Rhys's texts explore a whole series of limit-spaces, between self and other, home and abroad, interiority and exteriority, and metropolis and imperial margin. Her depiction of these liminal sites, such as the unstable Carib Territory, is characterised primarily by a movement through and between those spaces, and it is this heterotopic shifting that locates her work firmly within modernist images of the urban experience as well as the postcolonial experience of exile and displacement.[49]

Rhys described *Voyage* as having 'something to do with time being an illusion I think. I mean that the past exists – side by side with the present, not behind it; that what was – is.'[50] Problems with conventional notions of time mean that, instead of plain linear constructions, Rhys's texts often employ narratives structured spatially. Following Said, we can understand this spatialising aesthetic as conditioned by the imperial geography of Rhys's background. For instance, in *After Leaving Mr Mackenzie* (1930), when Julia returns to London from Paris the chiming of a church clock makes her feel that time has stood still: 'She felt her life had moved in a circle ... she had returned to her starting point, in this little Bloomsbury bedroom that was so exactly like the little Bloomsbury bedroom she had left nearly ten years before'.[51] The spatial trope of the circle, also used in 'Temps Perdi', thus displaces the temporal development of Julia's character. In *After Leaving Mr Mackenzie* Julia looks at a picture of a man encircled by a large corkscrew with the caption: *'La vie est un spiral, flottant dans l'espace'*.[52] This is echoed in Rachel Bowlby's reading of *Good Morning, Midnight*, where, Bowlby argues, the heroine Sasha is unable to differentiate between discrete times, between the past, the present and the future. Bowlby notes that this parallels a uniformity of places: all hotel rooms, streets and cities are the same for her, part of what Bowlby diagnoses as a fundamental impasse, both narrative and psychic, in the novel.[53]

Arguably *Voyage*, composed some years earlier, is not quite structured around this notion of narrative impasse, even though the same monotony of rooms and streets is ritualistically described. Here the voyage is very much the dominant metaphor, a journey through the byways and highways of Anna's mind, where the past is her vivid life in Dominica, and the present is her dulled existence in England. At one point Anna starts to count the towns she has visited in England while on tour as a dancer, and recalls how the bedrooms in the lodgings she has slept in were all 'exactly alike', with 'a high, dark wardrobe and something dirty red in the room' (*VD*, p. 128). This dreary image of interior space is then replaced by a long and vibrant reminiscence of a road leading to the 'Constance Estate' in Dominica. Here the 'feeling of the hills comes to you' and the scenery is alive, for there 'is never one moment of stillness – always something buzzing' (*VD*, p. 129). The novel sometimes exhibits to the reader a confusing movement through such different spaces, evoking Anna's own bewilderment at the spiral of life. Anna's mind shifts too much between past and present, London and the Caribbean, and she is unable to develop any identification of place in her memory. For Anna, therefore, the past becomes as real as the present, not because her life lacks change, but because she cannot fit the different parts of her life together, so that 'what was – is'. It is her psychic space that is in flux, being inhabited and overwhelmed by the vivacious imagery of other past lives and places, and unable to draw limits around these locations. This inability to demarcate borders again indicates Rhys's resistance to her colonial origins, where mapping the boundaries of territories became a priority of Empire. In Lefebvre's terms, Rhys's fiction always seems to offer a 'representational space' that contests the dominant imperial 'representation of space'.

Many of Anna's memories have the quality of what Freud calls 'screen memories', memories of childhood that are in fact only displaced versions of the real childhood memories that have been more deeply repressed. Freud questions whether 'we have any memories at all from our childhood: memories relating to our childhood may be all that we possess'.[54] Freud's theory suggests a primary lack at the heart of our memories, a psychic space to which we can never directly return, much like Rhys's futile attempt to rediscover the Imperial Road. Freud's idea captures something of Anna's state in *Voyage in the Dark*. The novel opens with the image of a stark break within herself: 'It was as if a curtain had fallen, hiding everything I had ever known. It was almost like being born again. The colours were different, the smells different, the feeling things gave you right down inside yourself was different' (*VD*, p. 7). This psychic splitting is brought about by the contrast between Anna's experience of the cold, grey monotony of the English city and the sensual images of (unnamed) Dominica: 'Sometimes it was as if I were back there and as if England were a dream. At other times England was the real thing and out there was the dream, but I could never fit them together' (*VD*, p. 8). These sorts of divisions – the 'Two Tunes' of the original title of the novel – run through the book, and are figured in a number of other spatial images: the dark high wall that Anna cannot get over or past, or the numerous mirrors into which she gazes, commenting on how she seems to be 'looking at somebody else' (*VD*, p. 21).

Anna is unable to fit these divided images together, but the more significant point is that the blurring of the boundaries around such images illustrates the disorientating power of modernity. Even in the reference to her inability to distinguish whether England or Dominica is a dream – that quality of being a screen – we discern a slippage between the inner psychic space of memory and the external space of a country: 'out there' is the rather puzzling term used for Dominica, as if the island exists as a location just outside Anna's room. After one incident, where she looks in a mirror and then lies down on a bed in a room that has a 'secret feeling', we read: 'I felt as if I had gone out of myself, as if I were in a dream' (*VD*, 21). Her inner sense of self is externalised, as she becomes merely an empty room waiting for her lover Walter to return. This switching between different spatial locations is the key to Rhys's encounter with modernity, an encounter that is simultaneously about a personal space of memory, a gendered experience of particular spaces (such as that of the hotel rooms where her relationships occur), and the imprint of the history of imperialism. The supreme pleasure of Rhys's work is just this fictional representation of movement between diverse spaces, so that to privilege any one space – imperialism, memory, gender – is to neglect the most significant feature of all: their interplay.

Anna's sensations of self, other and elsewhere prefigure an argument made by Richard Sennett, one of the most distinctive of urban critics, in *The Conscience of the Eye* (1990). In a discussion of James Baldwin, Sennett comments on how, in the modern city, one speaks to those who are Other, culturally different, as part of a continual process of attempted connection to the social world outside the atomised urban self: 'to sense the Other, one must do the work of accepting oneself as incomplete'.[55] This experience, whereby a person becomes 'like a foreigner to him or herself', is described by Sennett as a 'non-linear experience of difference', when the 'incomplete self' must turn outwards towards the Other in order to progress in understanding. This experience, notes Sennett, 'might be thought of as an *émigration extérieure*. One goes to the edge of oneself.

But precisely at that edge, one cannot represent oneself to oneself. Instead one sees, talks, or thinks about what is outside, beyond the boundary.'[56]

Anna's experiences share something of this quality of trying to cross the limits of her 'incomplete self', but her attempt is figured in terms slightly different from those of Sennett's account. For Rhys what lies outside the boundary of the self is the material nexus of rooms, streets and cafés in the city, rather than her relations with other people. As Julia, in *After Leaving Mr Mackenzie*, puts it: 'It was always places that she thought of, not people.'[57] The specific difficulty staged in Rhys's work is that the places to which her characters are attracted do not function as sites of identification sufficient to complete their 'incomplete selves'. Rooms themselves often seem alive to Anna, but in a threatening fashion, unlike Bachelard's benign sense of 'intimate space'.[58] When Anna's lover Walter takes her to his room in Green Street, in fashionable Mayfair, she feels that the room casts a spiteful eye over her: 'the rest of the house dark and quiet and not friendly to me. Sneering faintly, sneering discreetly, as a servant would. Who's this? Where on earth did he pick her up?' (*VD*, p. 43) In another room in Walter's house Anna is again sneered at, this time by a bust of Voltaire (*VD*, p. 75). Anna feels more comfortable only in rooms in the cheaper areas of London, such as the lodgings near the Chalk Farm Underground station whose decor reminds her of a restaurant (*VD*, p. 35). The importance of rooms to Rhys's protagonists is also shown by the fact that descriptions of hotel rooms form the first images of two of her other novels, *After Leaving Mr Mackenzie* and *Good Morning, Midnight*. In the latter, it is the room itself that speaks the novel's first words: '"Quite like old times," the room says. "Yes? No?"'[59]

Although the use of first-person narration in *Voyage* entails that Anna's mind is the primary lens through which we understand other spaces – rooms, streets, England, Dominica – the text often makes it difficult to decide from where Anna starts to comprehend her surroundings. Often the active terms of location are inanimate objects such as rooms and streets. Anna, like other Rhys characters, paradoxically experiences both claustrophobia and agoraphobia (*VD*, p. 120) in the novel, fearing rooms bearing down upon her or streets that might engulf her (*VD*, p.151). Her friend Germaine notes that England is 'a very nice place ... as long as you don't suffer from claustrophobia' (*VD*, p.70). Both phobias seem more than just projections of her anxieties, as the following passage shows:

> Then the taxi came; and the houses on either side of the street were small and dark and then they were big and dark but all exactly alike.

> And I say that all my life I had known that this was going to happen, and that I'd been afraid for a long time. There's fear, of course, with everybody. But now it had grown, it had grown gigantic; it filled me and it filled the whole world. (*VD*, p. 82)

Fear takes on the expansive qualities of the urban environment here, becoming dark and intimidating, blurring the line between inner space and external reality. Emotions themselves take on spatial characteristics and threaten to overwhelm the fragile space of Anna's selfhood. A similar description of houses as threatening is found in *Good Morning, Midnight*: 'Walking in the night with the dark houses over you, like monsters ... Tall cubes of darkness, with two lighted eyes at the top to sneer.'[60] And in *After Leaving Mr Mackenzie* Julia also feels the dark streets of London to be oppressive: 'It was the heavy darkness, greasy and compelling. It made walls round you, and shut you in so that you felt you could not breathe.'[61] These 'spatial phobias of modernity' demonstrate how Rhys's protagonists find de Certeau's stable 'identification of place' impossible to achieve.

Having been called 'a tart' by her landlady, Anna recalls a story 'about the walls of a room getting smaller and smaller until they crush you to death' (*VD*, p. 26). This nightmare of Bachelardian 'intimate space' rapidly shifts outdoors: '"I believe this damned room's getting smaller and smaller," I thought. And about the rows of houses outside, gimcrack, rotten-looking, and all exactly alike' (*VD*, p. 26). To escape the feeling of claustrophobia Anna's consciousness voyages outside the confines of the room, desperate for a setting in which to relax and to assuage her dislike of being called a tart. Then she takes some quinine for her nausea, and the narrative swings rapidly between England and Dominica. First, she thinks: 'This is England, and I'm in a nice, clean English room with all the dirt swept under the bed' (*VD*, p. 27). Intimate space merges into national space and, again, she cannot settle within either. Anna's mind drifts back to a time when she was ill at home in Dominica, being attended to by the black servant Francine:

> She [Francine] changed the bandage round my head and it was ice-cold and she started fanning me with a palm-leaf fan. And then night outside and voices of people passing in the street – the forlorn sound of voices, thin and sad. And the heat pressing down on you as if it were something alive. I wanted to be black. I always wanted to be black ... Being black is warm and gay, being white is cold and sad. (*VD*, p. 27)

Even the journey back to Dominica is fraught, showing Anna unable to attach her identity to any neutral location. The text has traversed many different spaces, registering various emotions: the claustrophobia of dingy rooms and dull London streets exemplifies the monotony and constrictions of her life, impelling the desire for difference, for a different space, and for a different identity, here her desire to be 'racially' other in order to be happy. Now we are within the flux of modern experience diagnosed and drawn back from by someone such as Forster. Arguably, Woolf and Joyce do not represent this experience as unnerving or overwhelming, unlike Rhys, whose skill is to represent from within this radical shifting of perspectives. The fragmentary, drifting style of writing here, with the use of the connective 'And' to open sentences, only emphasises the sense of a movement through space. Though colonial space is the end-point of this particular narrative journey it is clearly another site of non-existent origins: Anna's identification with Francine is a phantasmatic desire for a supposed authenticity and belonging signified by blackness. But it is – like the origins of our childhood screen memories or the Imperial Road – always absent, and it is such absence that is so often present in Rhys's depiction of the city, fuelling the search for identity itself.

When Rhys returned to Dominica, in 1936, she wrote that she felt 'at home and not at home'.[62] And in a later story, 'The day they burned the books', an Englishman infects the Rhys-like narrator with 'doubts about "home", meaning England', by saying to her: '"You're not English; you're a horrid colonial."'[63] To which she replies: '"Well, I don't much want to be English ... It's much more fun to be French or Spanish or something like that."'[64] The security of home can never be reached; instead Anna must shuttle between different geographical projections of her identity. This trajectory is matched in the narrative's switching between the scenes set in London and Anna's recollections of Dominica. Even in the desire to be black the historical consequences of Dominica's colonial history intervene to complicate Anna's projected identification. She wants to be black but is unsure whether the black people she wants to identify with are the African slaves – the 18-year-old slave girl Maillotte Boyd whom she abjectly remembers while she loses her virginity to Walter – or the indigenous Caribs who are now 'practically exterminated' (*VD*, p. 91).

Anna's position is perhaps best described by Stuart Hall, in an essay in which he defines Caribbean identity as possessing 'a conception of "identity" which lives with and through, not despite, difference; by hybridity'.[65] When Anna meets her racist step-mother Hester, who suggests that Anna is possibly of 'mixed-race', she notices and becomes obsessed with an advertisement in a newspaper: 'What is Purity? For Thirty Thousand Years the Answer has been Bourne's Cocoa' (*VD*, p. 50). Although 'purity' may be taken to refer at once to her sexual purity and her 'fallen' morals, the fact that later we learn her father owned a cocoa plantation makes us read the purity image in terms of her 'racial' identity. 'I'm a real West Indian' (*VD*, p. 47), she has just said to Walter, and indeed she is, in a sense, because of her hybridised spatial and cultural identity. Somewhat like Woolf's *Orlando*, the many facets of Anna's identity can never be synthesised into a singular identity defined by one place: her existence is anxiously held within the vertiginous spaces of modernity – psychic, intimate, national and abroad. In this sense Anna's identity is not constituted by connection, but by movement. As Stuart Hall comments of one component of Caribbean identity, that of the lost new world of indigenous peoples such as the Caribs of Dominica:

> It is because this New World is constituted for us as a place, a narrative of displacement, that it gives rise so profoundly to a certain imaginary plenitude, recreating the endless desire to return to 'lost origins', to be one again with the mother, to go back to the beginning ... And yet this 'return to the beginning' is like the imaginary in Lacan – it can neither be fulfilled nor requited, and hence is the beginning of the symbolic, of representation, infinitely renewable source of desire, memory, myth, search, discovery.[66]

Perhaps the endless quality of this search for identity illuminates the conclusion of *Voyage in the Dark*, where Anna, recovering from an abortion, hears the doctor say that she will soon be able to 'start all over again in no time' and dwells upon these words: 'And about starting all over again, all over again ...' (*VD*, p. 159). Anna's difficulty lies in accepting this provisional notion of identity. Rhys herself had written of her desire as a child to be absorbed into the very earth of Dominica: 'to identify with it, or lose myself in it ... The earth was like a magnet which pulled me and sometimes I came near it, this identification or annihilation that I longed for.'[67] The end of *Voyage* shows the impossibility of Anna's identification with any specific place, the novel's original ending, where she dies after the abortion, having her embrace annihilation. In that sense Anna Morgan's fate typifies many modernist representations of the city: unable to find fixity, but also failing to accommodate to flux and spatial disorientation. Rhys's portrayal of modernist anomie, however, is complexified by her relationship to the historical geography of imperialism: in Anna's sense of having to 'start all over again' we witness the same motif of circularity and flight noted in 'Temps Perdi'. In *Voyage in the Dark* we see not lost or wasted time, but the lost space of origins occupied, for Anna Morgan, by the black people with whom she wishes, impossibly, to identify. It is, however, this absent origin that, as Hall notes, impels representation and Rhys's fictional traversal of different spaces.

Contested topographies

In his 'Preface' to Jean Rhys's first publication, the book of short stories entitled *The Left Bank* (1927), Ford Madox Ford noted the correlation in her work of psychic experience and material spaces:

> I tried ... very hard to induce the author of *The Left Bank* to introduce some sort of topography of that region ... But would she do it? No! With cold deliberation, once her attention was called to the matter, she eliminated even such two or three words of descriptive matter as had crept into her work. Her business was with passion, hardship, emotions: the locality in which these things are endured is immaterial. So she hands you the Antilles with its sea and sky – 'the loveliest, deepest sea in the world – the Caribbean!' – the effects of landscape on the emotions and passions of a child being so penetrative, but lets Montparnasse, or London, or Vienna go. She is probably right. Something human should, indeed, be dearer to one than all the topographies of the world.[68]

Deborah Parsons has shown how mistaken Ford was in this judgement, in the sense that he was unable to see the precise fashion in which Rhys handles the relation between emotion and landscape, and the manner in which she does produce a topography.[69] Rhys contests Ford's topography by offering her own version; rejecting Ford's bohemia to demonstrate how difficult it is for a woman to fit into such urban spaces. As Shari Benstock points out, the Paris *rive gauche* portrayed by Rhys is not that of the 'Latin Quarter', but of the thirteenth and fourteenth *arondissements* further to the south-east of the city, a region more shabby and marginal than the cultural centre of Ford's literary geography.[70] In *Quartet*, for example, Marya dislikes the Boulevard St Michel with its 'glaring cafés', preferring the more 'dimly lit' Boulevard Montparnasse that leads away from the Latin Quarter.[71] Rhys's characters live on the edge of the bohemian city, a site replicating their marginal status as single women, and Rhys's status as colonial émigrée to the metropolis.

In de Certeau's terms, then, Rhys offers urban tours rather than a map of the city as desired by Ford. The city does not escape representation;

rather, Rhys's mode of urban description is always filtered through other spaces, notably that of abroad. In *Voyage in the Dark* she writes in and out of different spaces, shifting between inner, outer and abroad in an attempt to capture a set of emotional responses to her perception of modernity. She rejected producing a one-dimensional literary map of cities such as Paris or London because her spatial history impelled her to understand the metropolis in relation to the imperial margin. We witness in her texts, then, a voyage through and into modernity, a quest for identity without origin or end, from imperial metropolis to dream-like colony and back again, starting all over again. Here we have, as some recent critics have begun to notice, one reason for the current revival of interest in Rhys's work. For in its attention to provisionality, flux and hybridity in terms of cultural identity and belonging it prefigures certain key motifs of postmodernist thought. Helen Carr, for example, suggests that 'Rhys's fiction registers the sense of disorientation and the uncertain identity of those who live the ambivalent, uncentred, dislocated existences which some now argue have become paradigmatic of our postmodernist times.'[72] Whereas certain modernist writers find this prospect – what I have been calling the idea of 'flux' – deeply troubling and one which must, if possible, be contained by notions of form, pattern or myth,[73] I think that Rhys's work shows an awareness that this mobility cannot be avoided or simply contained by modernist form. It also demonstrates how 'flux' or 'mobility' are experiences of space that are historically and socially produced, here by the intertwined histories of Dominica and Europe.[74]

The difficulty of containing this flux within the formal spaces of the literary text is shown in one of Anna's earliest impressions of arriving in England. This incident also demonstrates Rhys's complex sense of topography, and a form of Lefebvrean understanding of social space. Anna can convert neither her new home in London nor her past home in Dominica into a place of identification:

> Lying between 15° 10′ and 15° 40′ N. and 61° 14′ and 61° 30′ W. 'A goodly island and something highland, but all overgrown with woods,' that book said. And all crumpled into hills and mountains as you would crumple a piece of paper in your hand – rounded green hills and sharply-cut mountains.
>
> A curtain fell and then I was here.
>
> ... This is England Hester said and I watched it through the train-window divided into squares like pocket-handkerchiefs; a small tidy look it had everywhere fenced off from everywhere else – what are those

> things – those are haystacks – oh are those haystacks – I had read about England ever since I could read – smaller and meaner everything is never mind – this is London – hundreds thousands of white people white people rushing along and the dark houses all alike frowning down one after the other all alike all stuck together – the streets like smooth shut-in ravines and dark houses frowning down – oh I'm not going to like this place I'm not going to like this place I'm not going to like this place – you'll get used to it Hester kept saying I expect you feel like a fish out of water but you'll soon get used to it ... (*VD*, pp. 15–16)

Anna cannot adequately grasp either national space. Dominica is now only an abstract geographical location, a Lefebvrean 'representation of space' conceptualised by means of longitude and latitude. The quotation Anna remembers is another such representation of space, a description in a book viewing the island through the colonial gaze, as it is taken from the island's first conqueror, Christopher Columbus, in his report on Dominica to the King of Spain in 1493.[75] Anna contrasts this textual image with one of her own, comparing the mountains of Dominica to a crumpled piece of paper. This image, however, is curtailed in the short sentence that follows, placed on a single line to capture the sharp division between the two locations. The voyage across the ocean transforms into Anna's journey by train into the metropolis, and the mobility is continued in the style of this paragraph describing her initial perception of England. The passage is a stream of impressions split by textual punctuation such as dashes that emphasise the spatial distance between each group of words. Opposed to the conceptual 'representations of space' in the previous paragraph, this is an example of Lefebvrean 'representational space', the 'lived' space of actual inhabitants trying to make sense of their location (*PS*, p. 39).

If Dominica can only be grasped in this detached fashion, London is now a place of foreboding, of dark 'ravines' and claustrophobic streets that can, so far, be only tolerated. England is a rigidly ordered set of spaces, with everything 'fenced off' or 'divided into squares'. The references to the darkness of the streets and houses indicate the trauma of Anna's voyage to the 'mother country' and the interconnected imperial histories of London and Dominica. Though London is, unlike Dominica, full of 'white people', her overwhelming impression of it is one of darkness. Discussing the symbolic value of darkness in relation to colonial discourse, Homi Bhabha argues for the profound ambivalence of the image: 'Darkness signifies at once both birth and death; it is in all cases a desire to return to the fullness of the mother.'[76] Anna's experience illustrates this point, as her association with the 'white' mother country is simultaneously a loss of her Dominican past, and an awareness of the deathly relationship between the two places. Dominica as a place has symbolically died, shown in the way that she can now only conceive of it abstractly in terms of the map or the report by Columbus.[77] Throughout the novel her memory will constantly return to this place, to keep touring through its spaces, hoping, impossibly, to be located there once again.

The city becomes the site where limits – between inside and outside, self and other, native and stranger – are transgressed, shown here in the stylistic space of the narrative with its rush of impressions and its disconnected, paratactic, structure. Cartographic representations of space are always problematical for Rhys's heroines. In one dismal lodgings in an English town Anna recalls another, similar, place where 'all the houses outside in the street were the same – all alike, all hideously stuck together – and the streets going north, east, south, west, all exactly the same' (*VD*, p.89).[78] In *Good Morning, Midnight* Sasha Jensen has a job as a tour guide in Paris, but becomes lost at the Place de l'Opéra: 'North, south, east, west – they have no meaning for me.'[79] Adjusting to this lack of orientation is a central theme of Rhys's fiction and it is why the spatial flux of the city is such an important image in much of her work in the 1920s and 1930s. As Rhys comments of Marya in *Quartet* (1928), it was because 'she was used to a lack of solidity and of fixed backgrounds'[80] that she did not find the mêlée of Paris particularly alarming.

Late in her life Rhys noted of some of the major places she had lived: 'There's Dominica, and that's completely separate from London, and then there's Vienna and Paris. My obsession is to fit it all together.'[81] Rhys attempts to fit together these different locations in the narrative space of her fiction, but too often, as with Anna's experience in *Voyage in the Dark*, the jigsaw never quite fits and we see the employment of a different trope, of a flight between pieces or places. Rhys's novels appear engaged in a difficult exploration across the spaces of modernism, to be somewhat inside rather than outside the 'flux', as it were, attempting to dwell and find 'home' in this vertiginous environment.

This desire to try to combine certain pieces of what Parsons calls Rhys's 'female psycho-geograph[y]'[82] also helps illuminate perhaps the most noticeable spatial feature of Rhys's texts: their peculiarly fragmented and discrete appearance on the page. Visually, the pages of Rhys's novels seem full of blank space, an effect reinforced by her tendency to use numerous single-sentence paragraphs and multiple sections to chapters. A related feature of her style is the constant use of line spaces or gaps between passages of prose. The effect of this heterotopic writing is thus to render her texts as mosaics, containing piecemeal images of her characters' lives and which match the disjointed geographies of Rhys's own life. It is perhaps most dramatically used in Anna's interior monologue, composed by Rhys as the original ending to *Voyage in the Dark*. The monologue giddily interweaves Anna's present sensations after an abortion with memories of her Caribbean childhood, concluding with her death. Over several pages full-stops are scantily used, there are many single line paragraphs, and breaks between phrases are only indicated by fissures in the typography:

> Father said a pretty useful mask   watch it and a slobbering tongue comes out   do you know I believe the whole damned business is like that   don't you think the idea of a malevolent idiot at the back of everything is the only one that fits the facts.[83]

Such writing combines together Foucault's twin senses of heterotopia: a form of writing that disorders grammar and syntax, and a disorientating location or 'place without a place' (OS, p. 27).

Anna Morgan's sense of geographical belonging is perhaps summed up in a comment by Rhys on being in exile: 'I thought, am I an expatriate? Expatriate from where?'[84] Any place from which Rhys or Anna might be exiled, or to which she might return, does not exist in any unproblematical sense, whether it is Dominica or London. For these places are not themselves anchored, but exist only in the spatial flux of history itself. Often in Rhys's texts we get the impression that once a character crosses a particular border – the threshold of a room or that of a country – for example, the very idea of returning becomes problematical.[85] *Returning* implies that something remains the same to which one is restored, a feature absent both for Rhys's sense of belonging and for the city itself, since a city exists only in the process of reconstituting and refiguring itself. Said, discussing the transportation of Magwitch in Dickens's *Great Expectations*, comments tellingly on the relation between metropolitan centre and imperial dominions implied in this incident: 'subjects can be taken to places like Australia, but they cannot be allowed to "return" to metropolitan space'.[86] Anna Morgan is also subject to this interdiction. She is literally allowed to 'return' to England, the land of her supposed origins in terms of family, language and culture, and this difference perhaps marks one historical change in imperial–metropolitan relations by the early twentieth century. But the meanings of 'return' are still problematical, for Anna is perceived as profoundly not-English: she is called the 'Hottentot' by the other girls in her touring company and her non-standard accent is often the subject of comment (*VD*, p. 12). Rhys's depiction of the drama of return demonstrates the profound difficulty modernist writing had with the notion of home: alienation in the city, or the hostile, beliefless, modern world, seem only a metaphoric version of this spatial alienation from home articulated in Anna's return to the metropolitan homeland. Or as Rhys later commented of her own early experiences in London: 'I would never be part of anything. I would never really belong anywhere, and I knew it, and all my life would be the same, trying to belong, and failing ... I am a stranger and I always will be.'[87] 'Home' for Rhys becomes a kind of screen memory, in Freud's sense: a place to which one can only imagine returning. 'Trying to belong' becomes a voyage through the multiple metropolitan spaces of Europe, a rather labyrinthine tour with no serviceable map.

Here we reach an understanding of how a text like *Voyage in the Dark* might fit into a revised literary geography of modernism. For its depiction of metropolitan anomie is a kind of palimpsest of Rhys's experience as a colonial migrant, where being a stranger is the norm rather than the exception. Both are instances of what Homi Bhabha has recently characterised 'unhomely fictions' in the context of postcolonial and postmodernist literature, such as that of Toni Morrison and Nadine Gordimer.[88] In this discussion of Jean Rhys I have suggested that 'unhomely fictions' are first found in the writing of modernism, and that perhaps modernism itself might be considered an 'unhomely' phenomenon, in the sense that it explores, whether in the context of the European metropolis or in that of the colony, geographies that are always in flux and for which the idea of a settled place of origin does not exist. This spatial history of modernism needs further exploration; but this will not be a voyage in the dark so much as a journey across and between different spaces: psychic, domestic, national, international and textual.

Notes

1 Jean Rhys, *Wide Sargasso Sea* (Harmondsworth, Penguin, 1968), p. 148.

2 Jean Rhys, *Voyage in the Dark* (Harmondsworth, Penguin, 1969).

3 Rhys has perhaps received less attention than one might expect from feminist critics. On this point see Helen Carr, *Jean Rhys* (Plymouth, Northcote House, 1996), p. 12. Another counter-example is afforded by Sylvie Maurel, *Jean Rhys*, Women Writers Series (Basingstoke, Macmillan, 1998). She is, however, included in the revisionist modernist anthology edited by Bonnie Kime Scott, *The Gender of Modernism: A Critical Anthology* (Bloomington, Indiana University Press, 1990).

4 For a discussion of Heidegger's notion of dwelling see chapter 1, this book.
5 Jean Rhys, *Quartet* (Harmondsworth, Penguin, 1973), p. 10.
6 Walter Benjamin, *Charles Baudelaire: A Lyric Poet in the Era of High Capitalism*, trans. Harry Zohn (London, Verso, 1983), p. 37.
7 One exception to this is Mary Lou Emery, *Jean Rhys at 'World's End': Novels of Colonial and Sexual Exile* (Austin, University of Texas Press, 1990), who argues that the early novels set in Europe are 'experiments in female and Third World modernism written in dialogue with European modernism' (p. xiv). However, Emery's focus is not specifically on the geographical aspects of these modernisms. For an examination of how Rhys's work explores Caribbean culture see Elaine Savory, *Jean Rhys* (Cambridge, Cambridge University Press, 1998).
8 D. H. Lawrence, 'New Mexico' in *A Selection from Phoenix*, ed. A. A. H. Inglis (Harmondsworth, Penguin, 1971), p. 125. Paul Fussell argues that the 1930s was the last golden age of travel writing because of the onset of tourism: see Fussell *Abroad: British Literary Travelling between the Wars* (New York and Oxford, Oxford University Press, 1980) p. 41.
9 Claude Lévi-Strauss, *Tristes Tropiques*, trans. John and Doreen Weightman (Harmondsworth, Penguin, 1976), p. 50.
10 Fussell, *Abroad*, p. 39. This is a problematical distinction, as Caren Kaplan has shown, relying upon a quasi-imperialist rhetoric of exploration, where conquest masquerades as 'discovery': see Caren Kaplan, *Questions of Travel: Postmodern Discourses of Displacement* (Durham and London, Duke University Press, 1996), pp. 49–57.
11 For a consideration of the use of cartography in Greene's book see my 'Journey with maps: travel theory, geography and the syntax of space', in Charles Burdett and Derek Duncan, eds, *Cultural Encounters: European Travel Writing in the 1930s* (Oxford, Berghahn, forthcoming).
12 Rachel Bowlby, '"The impasse": Jean Rhys's *Good Morning, Midnight*' in *Still Crazy After All These Years: Women, Writing and Psychoanalysis* (London, Routledge, 1992), p. 53.
13 Jan Montefiore notes that there is a great deal of autobiographical writing by women in the 1930s that differs from the writings of men of the period in assuming a position of marginality and in a choice of fiction rather than memoir, as in travel writing. See Jan Montefiore, 'Case-histories versus the "undeliberate dream": men and women writing the self in the 1930s', in Gabrielle Griffin, ed., *Difference in View: Women and Modernism* (London, Taylor & Francis, 1994).
14 Jean Rhys, *Smile, Please: An Unfinished Autobiography* (Harmondsworth, Penguin, 1981), p. 165.
15 Deborah L. Parsons, *Streetwalking the Metropolis: Women, the City and Modernity* (Oxford, Oxford University Press, 2000), pp. 124–5.
16 Parsons, *Streetwalking the Metropolis*, p. 136.
17 Edward Said, *Culture and Imperialism* (London, Vintage, 1994), p. 227.

18 Said, *Culture and Imperialism*, p. 228.
19 Said, *Culture and Imperialism*, p. 229.
20 For Rhys's use of literary quotation and allusion see Helen Carr, *Jean Rhys* (Plymouth, Northcote House, 1996), pp. 8–10. I discuss Rhys's use of a modified form of circularity below.
21 Said, *Culture and Imperialism*, p. 229. For a critique of Said's aesthetic conception of spatiality see Laura Chrisman, 'Imperial space, imperial place: theories of empire and culture in Fredric Jameson, Edward Said and Gayatri Spivak', *New Formations* 34 (1998), 53–69.
22 See Joseph Frank, 'Spatial form in modern literature' (1945), reprinted in Frank, *The Widening Gyre: Crisis and Mastery in Modern Literature* (Bloomington and London, Indiana University Press, 1968).
23 Quoted Carole Angier, *Jean Rhys: Life and Work* (Harmondsworth, Penguin, 1992) p. 655. For a brief discussion of contemporary Dominica see Peter Hulme, 'Islands of enchantment: extracts from a Caribbean travel journey', *New Formations* 3 (winter 1987), 93.
24 Paul Carter, *The Road to Botany Bay: An Exercise in Spatial History* (London and Boston, MA, Faber, 1987), p. 15.
25 Most of the information about the history of Dominica in this chapter is taken from the following: Peter Hulme and Neil L. Whitehead, eds, *Wild Majesty: Encounters with Caribs from Columbus to the Present Day. An Anthology* (Oxford, Clarendon Press, 1992); James Anthony Froude, *The English in the West Indies, or the Bow of Ulysses* (London, Longmans, Green & Co., 1888); and Teresa O'Connor, *Jean Rhys: The West Indian Novels* (New York and London, New York University Press, 1986). For a general account of perceptions of American Indians see Robert F. Berkhofer Jr, *The White Man's Indian: Images of the American Indian from Columbus to the Present* (New York: Vintage, 1979).
26 Hulme and Whitehead, *Wild Majesty*, p. 260.
27 Hulme and Whitehead, *Wild Majesty*, pp. 264, 268–9.
28 Maggie Humm, *Border Traffic: Strategies of Contemporary Women Writers* (New York and Manchester, Manchester University Press, 1991), p. 74.
29 See, for example, the essays in Anne Godlewska and Neil Smith, eds, *Geography and Empire: Critical Studies in the History of Geography* (Oxford, Blackwell, 1994).
30 Said, *Culture and Imperialism*, p. 93.
31 Derek Gregory, *Geographical Imaginations* (Oxford, Blackwell, 1994), p. 46.
32 Quoted in David N. Livingstone, *The Geographical Tradition: Episodes in the History of a Contested Discipline* (Oxford, Blackwell, 1992), p. 219.
33 H. J. Mackinder, *Britain and the British Seas*, 2nd edn (Oxford, Clarendon Press, 1907), p. 4. For a fascinating reading of Mackinder's geography that relates its formal strategies to those of literary modernism see Chris GoGwilt, 'The geopolitical image: imperialism, anarchism, and the hypothesis of cul-

ture in the formation of geopolitics', *Modernism/Modernity* 5: 3 (September 1998), 49–70.

34 Mackinder, *Britain and the British Seas*, p. 343.

35 Mackinder, *Britain and the British Seas*, pp. 346, 352.

36 Jean Rhys, 'Temps Perdi', *Penguin Modern Stories 1*, ed. Judith Burnley (Harmondsworth, Penguin, 1969), p. 88.

37 Rhys, 'Temps Perdi', p. 81.

38 Rhys, 'Temps Perdi', p. 81.

39 An intended, but unwritten, novel was to be called *Wedding in the Carib Quarter*, and Rhys, in a note, expressed the oblique view that the 'feeling I have about the Caribs & the Carib Quarter is very old & very complicated ... When I try to explain the feeling I find I cannot or do not wish to': Angier, *Jean Rhys*, p. 33.

40 Howells makes the point that the name of the English house in 'Temps Perdi', Rolvenden, suggests the idea of revolving, as shown in the windings of memory in the narrative: see Coral Ann Howells, *Jean Rhys* (New York and London, Harvester Wheatsheaf, 1991), p. 142.

41 See Angier, *Jean Rhys*, p. 356.

42 See James Donald, 'This, here, now: imagining the modern city', in Sallie Westwood and John Williams, eds, *Imaging Cities: Scripts, Signs, Memories* (London and New York, Routledge, 1997).

43 For an excellent discussion of this point in relation to the British ordnance survey map of Ireland see Mary Hamer, 'Putting Ireland on the map', *Textual Practice* 3: 2 (summer 1989), 184–201.

44 Rhys, 'Pioneers, Oh, Pioneers' in *Sleep it Off, Lady* (London, Andre Deutsch, 1976), p. 19.

45 Teresa O'Connor, 'Jean Rhys, Paul Theroux, and the Imperial Road', *Twentieth Century Literature* 38: 4 (winter 1992), 410–11.

46 Patrick Williams, 'Simultaneous uncontemporaneities: theorising modernism and empire', in Howard J. Booth and Nigel Rigby, eds, *Modernism and Empire* (Manchester, Manchester University Press, 2000), p. 25.

47 Avril Horner and Sue Zlosnik, *Landscapes of Desire: Metaphors in Modern Women's Fiction* (New York and London: Harvester Wheatsheaf, 1990), p. 136, note that the macaronic quality of *Good Morning, Midnight* shows 'the voice of someone at home in both languages [French and English] but who has no real sense of worth or "home" in either country'. The link between experimental linguistic practices and gatherings of migrants in Western cities in the early twentieth century is explored in Raymond Williams, *The Politics of Modernism: Against the New Conformists* (London: Verso, 1989), pp. 45–6.

48 In a late letter, of 14 September 1959, Rhys writes: 'As far as I know I am white – but I have no country really now.' See Jean Rhys, *Letters 1931–1966*, ed. Francis Wyndham and Diana Melly (Harmondsworth, Penguin, 1985), p. 172.

49 For discussion of modernism and the urban experience see, among others, Bradbury, 'The cities of modernism', in Malcom Bradbury and James McFarlane, eds, *Modernism: A Guide to European Literature 1890–1930* (Harmondsworth, Penguin, 1976); Williams, *Politics of Modernism*; and the essays in E. Timms and D. Kelly, eds, *Unreal City: Urban Experience in Modern European Literature and Art* (Manchester, Manchester University Press, 1985).

50 Rhys, *Letters*, p. 24. Stressing time rather than space, Humm, *Border Traffic*, pp. 85–6, sees this aspect of the novel as part of Rhys's depiction of a 'women's maternal history' that stands outside of patriarchal 'chronological history'.

51 Jean Rhys, *After Leaving Mr Mackenzie* (Harmondsworth, Penguin, 1971), p. 48.

52 Rhys, *After Leaving Mr Mackenzie*, p. 13.

53 Bowlby, '"The impasse"', p. 36. Rhys's representations of Paris and London are, contra Bowlby, quite different, with Paris often perceived as more positive, especially in *After Leaving Mr Mackenzie*.

54 Sigmund Freud, 'Screen memories' (1899) in *Collected Papers*, ed. James Strachey (New York, Basic Books, 1959), vol. 5, p. 69.

55 Richard Sennett, *The Conscience of the Eye: The Design and Social Life of Cities* (London, Faber, 1990), p. 148.

56 Sennett, *Conscience of the Eye*, p. 148.

57 Rhys, *After Leaving Mr Mackenzie*, p. 9.

58 For a discussion of Bachelard's 'intimate space' see ch. 1, this book.

59 Rhys, *Good Morning, Midnight* (Harmondsworth, Penguin, 1969), p. 9.

60 Rhys, *Good Morning, Midnight*, p. 28.

61 Rhys, *After Leaving Mr Mackenzie*, p. 62.

62 Angier, *Jean Rhys*, p. 354.

63 Rhys, *Tigers Are Better Looking, with a Selection from The Left Bank* (Harmondsworth, Penguin, 1972), p. 38.

64 Rhys, *Tigers Are Better Looking*, p. 39.

65 Stuart Hall, 'Cultural identity and diaspora', in Patrick Williams and Laura Chrisman, eds, *Colonial Discourse and Postcolonial Theory: A Reader* (New York and London, Harvester Wheatsheaf, 1993), p. 402. Hybridity has become a deeply contested notion within postcolonial theory; for one brief critique see Anthony Easthope, 'Bhabha, hybridity and identity', *Textual Practice* 12: 2 (1998), 341–8.

66 Hall, 'Cultural identity and diaspora', p. 402.

67 Rhys, *Smile Please*, pp. 81–2.

68 Ford Madox Ford, 'Preface' to Jean Rhys, *The Left Bank and Other Stories* (London, Jonathan Cape, 1927).

69 Parsons, *Streetwalking the Metropolis*, pp. 137–9.

70 Shari Benstock, *Women of the Left Bank: Paris 1900–1940* (London, Virago, 1987), p. 449.

71 Rhys, *Quartet*, p. 54.

72 Carr, *Jean Rhys*, p. 23.

73 As, for example, in the attention to structuring devices of myth or order in modernist works such as T. S. Eliot's *The Waste Land* or James Joyce's *Ulysses*. On this point see Terry Eagleton, *The Ideology of the Aesthetic* (Oxford, Blackwell, 1990), pp. 316–25.
74 For a similar argument about the material basis of mobilities see Tim Cresswell, 'The production of mobilities', *New Formations* 43 (2001).
75 See O'Connor, *Rhys: The West Indian Novels*, p. 13.
76 Homi K. Bhabha, *The Location of Culture* (London and New York, Routledge, 1994), p. 82. The most thorough discussion of the psychoanalytic relevance of the mother in Rhys is found in Deborah Kelly Kloepfer, *The Unspeakable Mother: Forbidden Discourse in Jean Rhys and H. D.* (Ithaca, NY, and London, Cornell University Press, 1989).
77 In *Wide Sargasso Sea* it is England which is figured in terms of cartographic representation, as Antoinette thinks of 'England, rosy pink in the geography book map', with a set of information about exports, imports and the 'character of inhabitants' (p. 92).
78 As if to emphasise this undifferentiated experience of space Rhys has Anna repeat this phrase when she moves into Laurie's flat in London, towards the end of the novel (*VD*, p. 152).
79 Rhys, *Good Morning, Midnight*, p. 26.
80 Rhys, *Quartet*, p. 14.
81 See Angier, *Jean Rhys*, p. 631.
82 Parsons, *Streetwalking the Metropolis*, p. 132.
83 Rhys, '*Voyage in the Dark* part IV (original version)', reprinted in Bonnie Kime Scott, ed., *The Gender of Modernism: A Critical Anthology* (Bloomington, Indiana University Press, 1990), p. 385. Rhys's erratic use of punctuation is noticeable in the handwritten manuscript versions of many of her novels. In *After Leaving Mr Mackenzie*, for example, it is difficult to discern the paragraphing and there is often no use of speech marks; often passages are divided by a line of full-stops or by the use of gaps in the text. See Jean Rhys, *After Leaving Mr Mackenzie*, British Library, BL Add MS 57856.
84 See Howells, *Jean Rhys*, p. 20.
85 For a good account of the metaphorical role of rooms and borders in Rhys's work see Horner and Zlosnik, *Landscapes of Desire*, pp. 133–61.
86 Said, *Culture and Imperialism*, p. xvii.
87 Rhys, *Smile, Please*, p. 124.
88 Bhabha, *Location of Culture*, p. 9. Bhabha's further characterisation of this writing as exploring the 'inbetween spaces' (p. 1) of cultural identities can also be applied to Rhys. See Carr, *Jean Rhys*, pp. 32–3 for further discussion of this idea.

Conclusion: in lieu of a map

To conclude by situating Jean Rhys within debates around certain characteristics of postmodernism might seem perverse, or paradoxical at best, in a book devoted to modernism. However, this contextualisation emphasises once more that the bewildering postmodern hyperspace space diagnosed by Fredric Jameson has its origins in some of the experiences of space found in writers such as Rhys.[1] This is not to collapse together modernist and postmodernist conceptions of space, but rather to emphasise, contra Jameson, that space and geography were active and pervasive interests for modernist writers. Something akin to Homi Bhabha's 'unhomely fictions' is uncovered in the explorations of space and place in many modernist texts, where an understanding of geographical location in a rapidly changing material world appeared to be of crucial importance. As Ford Madox Ford noted, modernists too felt that they lived in 'spacious times'. Articulating more fully this spatial history of modernism, and tracing the many movements through the geographies of modernity represented by writers such as Forster, the Imagists, Joyce, Woolf and Rhys, has been what, hopefully, this book has achieved.

This book has the character, to borrow from de Certeau for one final time, of a short tour through modernism in Britain and Ireland, rather than a comprehensive map of modernism's engagement with the topics of space and geography. I hope that the various spatial stories of that engagement have become more prominent, encouraging a new awareness of how we might read other modernist texts in this vein. Clearly, in a book of this nature I could have analysed any number of other texts by my chosen writers in order to explore the geographical imaginings of modernism: Joyce's *Dubliners* represents a particularly rich topographical account of a city; the role of Paris in other Rhys novels merits much more consideration; Forster's 'Italian' novels (*Where Angels Fear to Tread* and *A Room with a View*) represent an intriguing exploration of 'Englishness', not to mention the colonial geography of *A Passage to India*; and Woolf's re-imagined geography of the Hebrides (via Cornwall) in *To the Lighthouse* could also be reconsidered.

Looking outside of the authors discussed here opens up a seemingly endless vista of other spatial visions of modernity. To take four, random, examples: I might have discussed Katherine Mansfield's New Zealand perspective on modernism; Dorothy Richardson's extended portrait of the metropolis of London in *Pilgrimage*; Gertrude Stein's spatial experiments in *Geography and Plays* (1922) and *The Geographical History of America* (1935); or D. H. Lawrence's search for a more Heideggerian place of dwelling, manifested throughout his works. But considerations of space and time have produced the itinerary which this book has taken in.

Another book might have emerged out of this series of tours through modernism and modernity, one concerned with how modernism engaged with the many technologies of transport. The development of new technologies is clearly a central feature of how modernity felt different to people living in the early twentieth century: the impact of electricity, for example, was far-reaching. There is much more that could be written about how technologies of transport signified a modernity of change, upheaval and possibility to modernist writers: from bicycles and the 'new woman' to the many images of aeroplanes in Woolf's fiction.[2] Another avenue might have been to explore the links between transport, modernism and travel writing. The future might hold such a book on modernism and the experience of transport. Rather than explore that fascinating theme, however, I have frequently returned to the cultural history of cars, buses and trains in the early years of the twentieth century in order to help interpret the notions of movement and mobility in as materialist a fashion as possible. Here I have suggested that the flux and disorientation evident in the experiments of many modernist writers may better be understood by situating them within a spatial history of transport in the period. The different perceptions of various forms of transport also help tease out conflicting attitudes towards modernity evident in the writers here discussed: from the grudging presence of the motorcar in *Howards End* to the dramatic driving of *Orlando*.

One central aim of the book was to delineate a vocabulary drawing upon concepts employed by theorists of space and geography for the interpretation of literary modernism. Obviously, a critical literary geography can be applied to other literary texts and periods. This is emerging elsewhere within contemporary English studies, particularly in the fields of postcolonial writing and in forms of eco-criticism. In a 1998 review of several books addressing the representation of space and geography, John Kerrigan noted 'how valuable the findings of a new literary geography could be'.[3] The main argument outlined here about any such methodology is that it requires a more elaborated and theoretically sophisticated understanding of what a critical literary geography might look like. To simply employ a series of metaphors of travel, geography or mapping is clearly not sufficient. I hope this book has demonstrated the need for literary studies to engage more closely with the theories of space, place and geography found in writers such as Lefebvre, de Certeau and Foucault.

Cyril Connolly reported a conversation he once had with James Joyce in which, after some erudite discussion, Joyce turned to Connolly and said: 'I am afraid I am more interested, Mr Connolly, in Dublin street names than in the riddle of the universe.'[4] Having completed this book, I think that Joyce had a point. At the very least, an investigation into such spaces and geographies should further elucidate some of the riddles of modernism.

Notes

1 See the discussion of Jameson in ch. 3.

2 For a discussion of the aeroplane in Woolf see Gillian Beer, 'The island and the aeroplane: the case of Virginia Woolf', in Homi K. Bhabha, ed., *Nation and Narration* (London, Routledge, 1990).

3 John Kerrigan, 'The country of the mind', *Times Literary Supplement*, 11 September 1998, 3–4.

4 Cyril Connolly, *Previous Convictions* (London, Hamish Hamilton, 1963), p. 271.

Select bibliography

Agnew, John A. and James S. Duncan, eds, *The Power of Place: Bringing Together Geographical and Sociological Imaginations* (Boston, MA, Unwin Hyman, 1989).

Agnew, John, David N. Livingstone, and Alisdair Rogers, eds, *Human Geography: An Essential Anthology* (Oxford, Blackwell, 1996).

Aldington, Richard, 'Modern poetry and the Imagists', *The Egoist* (1 June 1914), 202.

Aldington, Richard, 'The poetry of F. S. Flint', *The Egoist* (1 May 1915), 80–1.

Aldington, Richard, *The Complete Poems of Richard Aldington* (London, Allen Wingate, 1948).

Anderson, Perry, 'Modernity and revolution', *New Left Review* 144 (March–April 1984), 96–113.

Angier, Carole, *Jean Rhys: Life and Work* (Harmondsworth, Penguin, 1992).

Anon., 'Don'ts for holiday travellers', *The Railway and Travel Monthly* 1: 5 (September 1910), 395.

Apollonio, Umbro, ed., *Futurist Manifestos* (London, Thames & Hudson, 1973).

Attridge, Derek and Marjorie Howes, eds, *Semicolonial Joyce* (Cambridge, Cambridge University Press, 2000).

Augé, Marc, *Non-Places: Introduction to an Anthropology of Supermodernity*, trans. John Howe (London, Verso, 1995).

Bachelard, Gaston, *The Poetics of Space* (1958), trans. Maria Jolas (Boston, MA, Beacon Press, 1969).

Bailey, Peter, 'Adventures in space: Victorian railway erotics', unpublished paper given at the Social History Conference 'Culture and subculture', Belfast, January 2001.

Bakhtin, Mikhail, *The Dialogic Imagination: Four Essays*, ed. Michael Holquist, trans. Caryl Emerson and Michael Holquist (Austin, University of Texas Press, 1981).

Balshaw, Maria and Liam Kennedy, eds, *Urban Space and Representation* (London, Pluto, 2000).

Bauman, Nicola, *Morgan: A Biography of E. M. Forster* (London, Hodder & Stoughton, 1993).

Bazargan, Susan, 'Mapping Gibraltar: colonialism, time, and narrative in "Penelope"' in *Molly Blooms: A Polylogue on 'Penelope' and Cultural Studies* (Madison and London, University of Wisconsin Press, 1994).

Beer, Gillian, 'The island and the aeroplane', in Homi K. Bhabha, ed., *Nation and Narration* (London and New York, Routledge, 1990).

Beer, Gillian, *Virginia Woolf: The Common Ground* (Edinburgh, Edinburgh University Press, 1996).

Bell, David, Jon Binnie, Julia Cream and Gill Valentine, 'All hyped up and no place to go', *Gender, Place and Culture* 1 (1994), 31–47.

Benjamin, Walter, *Charles Baudelaire: A Lyric Poet in the Era of High Capitalism*, trans. Harry Zohn (London, Verso, 1983).

Benjamin, Walter, *The Arcades Project*, trans. Howard Eiland and Kevin McLaughlin (Cambridge, MA, Belknap-Harvard University Press, 1999).

Benstock, Shari, 'City spaces and women's places in Joyce's Dublin', in Bernard Benstock, ed., *James Joyce The Augmented Ninth* (New York, Syracuse University Press, 1988).

Benstock, Shari, *Women of the Left Bank: Paris 1900–1940* (London, Virago, 1987).

Berkhofer Jr, Robert F., *The White Man's Indian: Images of the American Indian from Columbus to the Present* (New York, Vintage, 1979).

Berman, Marshall, *All that Is Solid Melts into Air: The Experience of Modernity* (London and New York, Verso, 1983).

Berriman, Algernon E., *Motoring: An Introduction to the Car and the Art of Driving it* (London, Methuen, 1914).

Bhabha, Homi K., *The Location of Culture* (London and New York, Routledge, 1994).

Bird, Anthony, *Early Motor Cars* (London, George Unwin, 1967).

Bird, Jon, Barry Curtis, Tim Putnam, George Robertson and Lisa Tickner, eds, *Mapping the Futures: Local Cultures, Global Change* (London, Routledge, 1993).

Blamires, Harry, *The New Bloomsday Book* (London and New York, Routledge, 1988).

Bloom, Clive, *Literature and Culture in Modern Britain*, vol. 1: *1900–1929* (London, Longman, 1993).

Booth, Howard J. and Nigel Rigby, eds, *Modernism and Empire* (Manchester, Manchester University Press, 2000).

Bowlby, Rachel, *Just Looking: Consumer Culture in Dreiser, Gissing and Zola* (London and New York, Methuen, 1985).

Bowlby, Rachel, '"The impasse": Jean Rhys's *Good Morning, Midnight*' in *Still Crazy After All These Years: Women, Writing and Psychoanalysis* (London, Routledge, 1992).

Bowlby, Rachel, *Still Crazy After All These Years: Women, Writing and Psychoanalysis* (London, Routledge, 1992).

Bowlby, Rachel, *Feminist Destinations and Further Essays on Virginia Woolf* (Edinburgh, Edinburgh University Press, 1997).

Boyce, D. G., *Nationalism in Ireland*, 2nd edn (London, Routledge, 1991).

Bradbury, Malcolm and James McFarlane, eds, *Modernism: A Guide to European Literature 1890–1930* (Harmondsworth, Penguin, 1976).

Brantlinger, Patrick, '"The Bloomsbury fraction" versus war and empire', in Carola M. Kaplan and Anne B. Simpson eds, *Seeing Double: Revisioning Edwardian and Modernist Literature* (Basingstoke, Macmillan, 1996).

Brewster, Dorothy, *Virginia Woolf's London* (New York, New York University Press, 1960).

Bristow, Joseph, *Effeminate England: Homoerotic Writing after 1885* (Buckingham, Open University Press, 1995).

Brooker, Peter, *A Student's Guide to the Selected Poems of Ezra Pound* (London, Faber, 1979).

Brown, C. W., *ABC of Motoring* (London, Henry J. Drane, 1909).

Bruce, J. Graeme, *Tube Trains Under London: An Illustrated History* (London, London Transport, 1968).

Buci-Glucksmann, Christine, 'Catastrophic utopia: the feminine as allegory of the modern', *Representations* 14 (Spring 1986), 215–30.

Buck-Morss, Susan, 'The *flâneur*, the sandwichman and the whore: the politics of loitering', *New German Critique* 39 (1986), 99–140.

Budgen, Frank, *James Joyce and the Making of 'Ulysses'* (Bloomington, Indiana University Press, 1960).

Burnett, John, *A Social History of Housing 1815–1985*, 2nd edn (London, Routledge, 1986).

Buzard, James, *The Beaten Track: European Tourism, Literature and the Ways to 'Culture' 1800–1918* (Oxford, Clarendon Press, 1993).

Carey, John, *The Intellectuals and the Masses: Pride and Prejudice among the Literary Intelligentsia, 1880–1939* (London, Faber, 1992).

Carr, Helen, *Jean Rhys* (Plymouth, Northcote House, 1996)

Carter, Paul, *The Road to Botany Bay: An Essay in Spatial History* (London, Faber, 1987).

Casey, Edward S., *The Fate of Place: A Philosophical History* (Berkeley and Los Angeles, University of California Press, 1997).

Caughie, Pamela L., ed., *Virginia Woolf in the Age of Mechanical Reproduction* (New York and London, Garland, 2000).

Cheng, Vincent J., *Joyce, Race and Empire* (Cambridge, Cambridge University Press, 1995).

Chilton, Randolph and Carol Gibertson, 'Pound's "Metro" hokku': the evolution of an image', *Twentieth Century Literature* 36: 2 (1990), 225–36.

Chrisman, Laura, 'Imperial space, imperial place: theories of empire and culture in Fredric Jameson, Edward Said and Gayatri Spivak', *New Formations* 34 (1998), 53–69.

Cianci, Giovanni, 'Futurism and the English avant-garde: the early Pound between Imagism and Vorticism', *Arbeiten aus Anglistik und Amerikanistik* 1 (1981), 3–39.

Colls, Robert and Philip Dodd, eds, *Englishness: Politics and Culture 1880–1920* (London, Croom Helm, 1986).
Conrad, Joseph, *Heart of Darkness*, ed. Robert Kimbrough (New York, Norton, 1988).
Cork, Richard, *Vorticism and Abstract Art in the First Machine Age* (London and Bedford, Gordon Fraser Gallery, 1975), vol. 1.
Country Homes The Official Guide of the Metropolitan Railway comprising Residential Districts of Buckinghamshire, Hertfordshire, and Middlesex (London, Metropolitan Railways, 1909).
Craig, Maurice, *Dublin 1660–1860* (Dublin, Allen Figgis, 1980).
Cresswell, Tim, 'The production of mobilities', *New Formations* 43 (2001), 11–25.
Crosland, T. W. H., *The Suburbans* (London, John Long, 1905).
Daly, Mary E., *Dublin: The Deposed Capital. A Social and Economic History 1860–1914* (Cork, Cork University Press, 1984).
Davidson, John, *Ballads and Songs* (London, John Lane, 1894).
Davidson, John, *Fleet Street and Other Poems* (London, Grant Richards, 1909).
Day, John R., *The Story of London's Underground* (London, London Transport, 1963).
De Certeau, Michel, *The Practice of Everyday Life*, trans. Steven Randall (Berkeley, University of California Press, 1984).
De Man, Paul, 'Literary history and literary modernity' in *Blindness and Insight: Essays in the Rhetoric of Contemporary Criticism* (London, Methuen, 1983).
Deane, Seamus, 'Joyce the Irishman', in Derek Attridge, ed., *The Cambridge Companion to James Joyce* (Cambridge, Cambridge University Press, 1990).
Deleuze, Gilles, 'What is a *dispositif*?' in *Michel Foucault, Philosopher*, trans. Tim Armstrong (Hemel Hempstead, Harvester Wheatsheaf, 1992).
Derrida, Jacques, *Writing and Difference*, trans. Alan Bass (London, Routledge, 1978).
Di Battista, Maria and Lucy McDiarmid, eds, *High and Low Moderns: Literature and Culture 1889–1939* (Oxford, Oxford University Press, 1996).
Dickenson, Goldsworthy Lowes, 'Motoring', *The Independent Review* (January 1904), 578–89.
Dickenson, Goldsworthy Lowes, 'The motor tyranny', *The Independent Review* (Oct. 1906), 15–22.
Doane, Mary Ann, 'Film and masquerade: theorizing the male spectator', *Screen* 23: 3–4 (September–October 1982), 74–87.
Donald, James, 'This, here, now: imagining the modern city', in Sallie Westwood and John Williams, eds, *Imaging Cities: Scripts, Signs, Memories* (London and New York, Routledge, 1997).

Driver, Felix and David Gilbert, eds, *Imperial Cities: Landscape, Display and Identity* (Manchester and New York, Manchester University Press, 1999).
Duffy, Enda, *The Subaltern 'Ulysses'* (Minneapolis, Univeristy of Minnesota Press, 1994).
Duncan, James and Derek Gregory, eds, *Writes of Passage: Reading Travel Writing* (London, Routledge, 1999).
Eagleton, Terry, *The Ideology of the Aesthetic* (Oxford, Blackwell, 1990).
Easthope, Anthony, 'Bhabha, hybridity and identity', *Textual Practice* 12: 2 (1998), 341–8.
Ellmann, Maud, *The Poetics of Impersonality: T. S. Eliot and Ezra Pound* (Brighton, Harvester Press, 1987).
Ellmann, Richard, *James Joyce*, revised edn (Oxford, Oxford University Press, 1983).
Emery, Mary Lou, *Jean Rhys at 'World's End': Novels of Colonial and Sexual Exile* (Austin, University of Texas Press, 1990).
Emig, Rainer, *Modernism in Poetry: Motivations, Structures and Limits* (London, Longman, 1995).
Eysteinsson, Astradur, *The Concept of Modernism* (Ithaca, NY, and London, Cornell University Press, 1990).
Fletcher, John Gould, 'From Babel's Night', *Coterie* 4 (Easter 1920), 37.
Fletcher, John Gould, *Life Is My Song* (New York and Toronto, Farrar & Rinehart, 1937).
Flint, F. S., *Otherworld Cadences* (London, Poetry Bookshop, 1920).
Flint, Kate, 'Fictional suburbia', in Peter Humm, Paul Stigant and Peter Widdowson, eds, *Popular Fictions: Essays in Literature and History* (London and New York, Methuen, 1986).
Ford, Ford Madox, *The Soul of London* (1905), ed. Alan G. Hill (London, Everyman, 1995).
Ford, Ford Madox, 'Preface' to Jean Rhys, *The Left Bank and Other Stories* (London, Jonathan Cape, 1927).
Forster, E. M., *Collected Short Stories* (Harmondsworth, Penguin, 1954).
Forster, E. M., *The Manuscripts of 'Howards End'*, Abinger edition, vol. 4a, ed. Oliver Stallybrass (London, Edward Arnold, 1973).
Forster, E. M., *A Room with a View* (Harmondsworth, Penguin, 1978).
Forster, E. M., *Howards End*, ed. Oliver Stallybrass (Harmondsworth, Penguin, 1983).
Foster, R. F., *Modern Ireland 1600–1972* (London, Penguin, 1988).
Foucault, Michel, *The Order of Things: An Archaeology of the Human Sciences*, trans. A. M. Sheridan (London, Routledge, 1970).
Foucault, Michel, *Discipline and Punish: The Birth of the Prison*, trans. A. M. Sheridan-Smith (Harmondsworth, Penguin, 1977).
Foucault, Michel, *Power/Knowledge: Selected Interviews and Other Writings 1972–1977* (London, Harvester Wheatsheaf, 1980).

Foucault, Michel, *The Foucault Reader*, ed. Paul Rabinow (London, Penguin, 1984).

Foucault, Michel, 'Of other spaces' *Diacritics* 16: 1 (spring 1986), 22–7.

Foucault, Michel, 'Thought of the outside' in *Essential Works of Foucault*, vol. 2: *Aesthetics, Method and Epistemology*, trans. Robert Hurley and others (London, Allen Lane–Penguin, 1998).

Frank, Joseph, 'Spatial form in modern literature' (1945) in *The Widening Gyre: Crisis and Mastery in Modern Literature* (Bloomington and London, Indiana University Press, 1968).

Frank, Joseph, 'Spatial form: an answer to critics', *Critical Inquiry* 4 (1977–78), 231–52.

Frank, Joseph, 'Spatial form: some further reflections', *Critical Inquiry* 5 (1978–79), 275–90.

Freud, Sigmund, *Collected Papers*, 5 vols, ed. James Strachey (New York, Basic Books, 1959).

Freud, Sigmund, *Penguin Freud Library*, 15 vols, eds Angela Richards and Albert Dickson (Harmondsworth, Penguin, 1973–86).

Frickley, Pierrette M., ed., *Critical Perspectives on Jean Rhys* (Washington, DC, Three Continents Press, 1990).

Friel, Brian, *Translations* (London, Faber, 1981).

Froude, James Anthony, *The English in the West Indies, or the Bow of Ulysses* (London, Longmans, Green & Co., 1888).

Furbank, P. N., *E. M. Forster: A Life*, vol. 1: *The Growth of the Novelist 1879–1914* (London, Secker & Warburg, 1977).

Fussell, Paul, *Abroad: British Literary Travelling between the Wars* (Oxford, Oxford University Press, 1980).

Garvey, Johanna X. K., 'City limits: reading gender and urban spaces in *Ulysses*', *Twentieth Century Literature* 41: 1 (1995), 108–23.

Gibbons, Luke, 'Montage, modernism and the city', *Irish Review* 10 (1991), 1–6.

Gibbons, Luke, *Transformations in Irish Culture*, Critical Conditions: Field Day Essays (Cork, Cork University Press, 1996).

Gifford, Don, *'Ulysses' Annotated: Notes for James Joyce's 'Ulysses'*, 2nd edn (Berkeley, University of California Press, 1988).

Gilbert, Sandra M., 'Soldier's heart: literary men, literary women, and the Great War', in Elaine Showalter (ed.) *Speaking of Gender* (London and New York, Routledge, 1989).

Gilbert, Stuart, *James Joyce's 'Ulysses'* (1930) (New York, Vintage, 1955).

Gillies, John, *Shakespeare and the Geography of Difference* (Cambridge, Cambridge University Press, 1994).

Gilman, Charlotte Perkins, *The Charlotte Perkins Gilman Reader: 'The Yellow Wallpaper' and Other Fiction* (London, Women's Press, 1979).

Godlewska, Anne and Neil Smith, eds, *Geography and Empire: Critical Studies in the History of Geography* (Oxford, Blackwell, 1994).

GoGwilt, Chris, 'The geopolitical image: imperialism, anarchism, and the hypothesis of culture in the formation of geopolitics', *Modernism/ Modernity* 5: 3 (September 1998), 49–70.

GoGwilt, Chris, *The Invention of the West: Joseph Conrad and the Double-Mapping of Europe and Empire* (Stanford, CA, Stanford University Press, 1995).

Gordon, Lyndall, *Eliot's Early Years* (Oxford, Oxford University Press, 1977).

Granville, Gary, ed. *Divided City: Portrait of Dublin 1913* (Dublin, O'Brien Educational, 1978).

Green, Oliver, *The London Underground: An Illustrated History* (London, Ian Allen, 1987).

Green, Oliver, *Underground Art: London Transport Posters 1908 to the Present* (London, Studio Vista, 1990).

Gregory, Derek, *Geographical Imaginations* (Oxford, Blackwell, 1994).

Groden, Michael, *'Ulysses' in Progress* (Princeton, NJ, Princeton University Press, 1977).

Hall, Stuart, 'Cultural identity and diaspora', in Patrick Williams and Laura Chrisman, eds, *Colonial Discourse and Postcolonial Theory: A Reader* (New York and London, Harvester Wheatsheaf, 1993).

Hallberg, Robert von, 'Libertarian Imagism' *Modernism/Modernity* 2: 2 (1995), 63–79.

Hamer, Mary, 'Putting Ireland on the map' *Textual Practice* 3: 2 (summer 1989), 184–201.

Hampstead and Highgate Railway, *London's Latest Suburbs: An Illustrated Guide to the Residential Districts Reached by the Hampstead Tube* (London, Charing Cross, Hampstead and Highgate Railway, 1910).

Hanscombe, Gillian and Virginia L. Smyers, *Writing for Their Lives: The Modernist Women 1910–1940* (London, Women's Press, 1987).

Hardy, Thomas, *Tess of the D'Urbervilles* (Harmondsworth, Penguin, 1978).

Harley, J. B., 'Deconstructing the map', in John Agnew, David N. Livingstone and Alisdair Rogers, eds, *Human Geography: An Essential Anthology* (Oxford, Blackwell, 1996).

Harley, J. B., 'Maps, knowledge and power', in D. Cosgrove and S. Daniels, eds, *The Iconography of Landscape* (Cambridge, Cambridge University Press, 1988).

Harris, Jose, *Private Lives, Public Spirit: Britain 1870–1914* (Harmondsworth, Penguin, 1994).

Harvey, David, *The Condition of Postmodernity* (Oxford, Basil Blackwell, 1990).

Hayman, David, *'Ulysses': The Mechanics of Meaning* (Madison, University of Wisconsin Press, 1982).

Heaney, Seamus, 'The sense of place' in *Preoccupations: Selected Prose 1969–1978* (London, Faber, 1980).

Hearne, R. P., *Motoring* (London, Routledge, 1908).
Heidegger, Martin, 'Building, dwelling, thinking' (1951) in *Poetry, Language, Thought*, trans. Albert Hofstadter (New York, Harper & Row, 1971).
Henchy, Patrick, 'Nelson's Pillar', *Dublin Historical Record* 10: 2 (June–August 1948), 53–63.
Herr, Cheryl, *Joyce's Anatomy of Culture* (Urbana, University of Illinois Press, 1986).
Hetherington, Kevin, *The Badlands of Modernity: Heterotopia and Social Ordering* (London and New York, Routledge, 1997).
Hobsbawm, Eric, 'Labour in the great city', *New Left Review* 166 (November–December 1987), 38–47.
Holtz, William, 'Spatial form in modern literature: a reconsideration', *Critical Inquiry* 4 (1977–78), 271–83.
Horner, Avril and Sue Zlosnik, *Landscapes of Desire: Metaphors in Modern Women's Fiction* (New York and London: Harvester Wheatsheaf, 1990).
Howells, Coral Ann, *Jean Rhys* (New York and London, Harvester Wheatsheaf, 1991).
Howson, H. F., *London's Underground* (London, Ian Allen, 1986).
Hueffer (Ford), Ford Madox, 'Modern poetry', *The Thrush* 1: 1 (December 1909), 39–53.
Hulme, Peter, 'Islands of enchantment: extracts from a Caribbean travel journey', *New Formations* 3 (winter 1987), 81–95.
Hulme, Peter and Neil L. Whitehead, eds, *Wild Majesty: Encounters with Caribs from Columbus to the Present Day. An Anthology* (Oxford, Clarendon Press, 1992).
Hulme, T. E., *Further Speculations*, ed. S. Hynes (Minneapolis, University of Minnesota Press, 1955).
Hulme, T. E., *Speculations: Essays on Humanism and the Philosophy of Art*, ed. Herbert Read (London and New York, Routledge & Kegan Paul, 1987).
Humm, Maggie, *Border Traffic: Strategies of Contemporary Women Writers* (New York and Manchester, Manchester University Press, 1991).
Imagist Anthology 1930 (London, Chatto & Windus, 1930).
Jameson, Fredric, *The Political Unconscious: Narrative as a Socially Symbolic Act* (London, Methuen, 1981).
Jameson, Fredric, 'Postmodernism, or the cultural logic of late capitalism', *New Left Review* 146 (July–August 1984), 53–92.
Jameson, Fredric, *Modernism and Imperialism* (Derry, Field Day Pamphlet, 1988).
Jameson, Fredric, *Postmodernism, or the Cultural Logic of Late Capitalism* (London, Verso, 1991).
Jarvis, Brian, *Postmodern Cartographies: The Geographical Imagination in Contemporary American Culture* (London, Pluto, 1998).

Jones, A. R., *The Life and Opinions of Thomas Ernest Hulme* (London, Gollancz, 1960).
Jones, Peter, ed., *Imagist Poetry* (Harmondsworth, Penguin, 1972).
Joyce, James, *Letters of James Joyce*, 3 vols; vol 1 ed. Stuart Gilbert; vols 2 and 3 ed. Richard Ellmann (New York, Viking, 1957, 1966).
Joyce, James, *Dubliners* (London, Penguin, 1992).
Joyce, James, *Ulysses*, ed. Jeri Johnson (Oxford, Oxford University Press, 1993).
Joyce, James, *Occasional, Critical and Political Writing*, ed. Kevin Barry (Oxford, Oxford University Press, 2000).
Kaplan, Caren, *Questions of Travel: Postmodern Discourses of Displacement* (Durham and London, Duke University Press, 1996).
Kaplan, Carola and Anne B. Simpson, eds, *Seeing Double: Revisioning Edwardian and Modernist Literature* (Basingstoke, Macmillan, 1996).
Keating, Peter, 'Literature and the metropolis', in Anthony Sutcliffe, ed., *Metropolis 1890–1940* (London, Mansell, 1984).
Keating, Peter, *The Haunted Study: A Social History of the English Novel 1875–1914* (London, Fontana, 1991).
Keith, Michael and Steve Pile, eds, *Place and the Politics of Identity* (London, Routledge, 1993).
Kenner, Hugh, *The Mechanic Muse* (Oxford, Oxford University Press, 1987).
Kenner, Hugh, *The Pound Era* (Berkeley and Los Angeles, University of California Press, 1971).
Kermode, Frank, 'A reply to Joseph Frank', *Critical Inquiry* 4 (1977–78), 579–88.
Kern, Stephen, *The Culture of Time and Space 1880–1918* (Cambridge, MA, Harvard University Press, 1983).
Kerrigan, John 'The country of the mind', *Times Literary Supplement*, 11 September 1998, pp. 3–4.
Kiberd, Declan, *Inventing Ireland: The Literature of the Modern Nation* (London, Vintage, 1996).
Kilroy, Jim, '"Galleons of the streets": Irish trams', *History Ireland* 5: 2 (summer 1997), 42–7.
Kloepfer, Deborah Kelly, *The Unspeakable Mother: Forbidden Discourse in Jean Rhys and H. D.* (Ithaca, NY, and London, Cornell University Press, 1989).
Langland, Elizabeth, 'Gesturing towards an open space: gender, form and language in *Howards End*', in Jeremy Tambling, ed., *E. M. Forster: New Casebook* (Basingstoke, Macmillan, 1995).
Lawrence, D. H., 'New Mexico' in *A Selection from Phoenix*, ed. A. A. H. Inglis (Harmondsworth, Penguin, 1971).
Le Corbusier, *The City of Tomorrow and its Planning*, trans. F. Etchells (1925) (New York, Dover, 1987).

Ledger, Sally and Scott McCracken, eds, *Cultural Politics at the* Fin de Siècle (Cambridge, Cambridge University Press, 1995).

Lee, Hermione, *Virginia Woolf* (London, Vintage, 1997).

Lees, Andrew, *Cities Perceived: Urban Society in European and American Thought 1870–1940* (Manchester, Manchester University Press, 1985).

Lefebvre, Henri, *The Production of Space* (1974), trans. Donald Nicholson-Smith (Oxford, Blackwell, 1991).

Lefebvre, Henri, 'Reflections on the politics of space', *Antipode* 8: 2 (May 1976), 30–7.

Léger, Fernand, 'Contemporary achievements in painting' (1914), in Edward F. Fry, ed., *Cubism* (London, Thames & Hudson, 1966).

Lehan, Richard, *The City in Literature: An Intellectual and Cultural History* (Berkeley and London, University of California Press, 1998).

Lévi-Strauss, Claude, *Tristes Tropiques*, trans. John and Doreen Weightman (Harmondsworth, Penguin, 1976).

Lewis, Wyndham, *BLAST* 1 (1914), ed. Bradford Morrow, reprinted (Santa Rosa, CA, Black Sparrow Press, 1989).

Lewis, Wyndham, *Time and Western Man* (1927) (Boston, MA, Beacon Press, 1957).

Lewis, Wyndham, *Blasting and Bombardiering* (1937) (London, Calder & Boyars, 1967).

Liscombe, R. W., *William Wilkins 1778–1839* (Cambridge, Cambridge University Press, 1980).

Livingstone, David N., *The Geographical Tradition: Episodes in the History of a Contested Discipline* (Oxford, Blackwell, 1992).

Lucas, John, 'Discovering England: the view from the train', *Literature and History* 6: 2 (autumn 1997), 37–55.

Lukács, Georg, 'The ideology of modernism' in *Realism in Our Time: Literature and the Class Struggle*, trans. John and Necke Mander (New York, Harper & Row, 1971).

Lynch, Kevin, *The Image of the City* (Cambridge, MA, MIT Press, 1960).

McArthur, Murray, 'Rose of Castille/rows of cast steel: figural parallelism in *Ulysses*', *James Joyce Quarterly* 24: 4 (summer 1987), 411–22.

Macgregor Wise, J., *Exploring Technology and Social Space* (London, Sage, 1997).

McIntyre, John, 'Confining plots: modernist space in Joyce's "Hades"', Paper presented at the 'Modernism and the city' seminar, New Modernisms II Conference, University of Pennsylvania, October 2000.

Mackinder, H. J., *Britain and the British Seas*, 2nd edn (Oxford, Clarendon Press, 1907).

MacLoughlin, Adrian, *Guide to Historic Dublin* (Dublin, Gill & Macmillan, 1979).

McLaughlin, Joseph, *Writing the Urban Jungle: Reading Empire in London from Doyle to Eliot* (Charlottesville and London, University Press of Virginia, 2000).

Marcus, Laura, *Virginia Woolf* (Plymouth, Northcote House, 1997).

Marks, John, 'A new image of thought', *New Formations* 25 (summer 1995), 66–76.

Massey, Doreen, 'Flexible sexism', *Environment and Planning D: Society and Space* 9 (1991), 31–57.

Massey, Doreen, 'A place called home', *New Formations* 17 (summer 1992), 3–15.

Massey, Doreen and Pat Jess, eds, *A Place in the World? Places, Cultures and Globalization* (Oxford, Oxford University Press, 1995).

Masterman, C. F. G., *The Condition of England* (London, Methuen, 1909).

May, Brian, *The Modernist as Pragmatist: E. M. Forster and the Fate of Liberalism* (Columbia and London, University of Missouri Press, 1997).

Meisel, Perry, *The Myth of the Modern: A Study in British Literature and Criticism after 1850* (New Haven, CT, and London, Yale University Press, 1987).

Minow-Pinkney, Makiko, 'Flânerie by motor car?', in Laura Davies and Jeanette McVicker, eds, *Virginia Woolf and Her Influences: Selected Papers from the 7th Annual Conference on Virginia Woolf* (New York, Pace University Press, 1998).

Minow-Pinkney, Makiko, 'Virginia Woolf and the age of motor cars', in Pamela L. Caughie, ed., *Virginia Woolf in the Age of Mechanical Reproduction* (New York and London, Garland, 2000).

Mitchell, W. J. T., 'Spatial form in literature: towards a general theory', *Critical Inquiry* 6 (1979–80), 539–67.

Montifiore, Jan, 'Case-histories versus the "undeliberate dream": men and women writing the self in the 1930s', in Gabrielle Griffin, ed., *Difference in View: Women and Modernism* (London, Taylor & Francis, 1994).

Moretti, Franco, *Signs Taken for Wonders: Essays on the Sociology of Literary Forms*, trans. Susan Fischer, David Forgacs and David Miller (London, Verso, 1983).

Moretti, Franco, *Atlas of the European Novel 1800–1900* (London and New York, Verso, 1998).

Morrisson, Mark S., *The Public Face of Modernism: Little Magazines, Audiences, and Reception 1905–1920* (Madison, University of Wisconsin Press, 2001).

Mulvey, Laura, *Visual and Other Pleasures* (London, Macmillan, 1989).

Mumford, Lewis, *The City in History: Its Origins, its Transformations and its Prospects* (Harmondsworth, Penguin, 1966).

Nairn, Ian and Nikolaus Pevsner, *The Buildings of England: Sussex* (Harmondsworth, Penguin, 1965).

Nava, Mica and Alan O'Shea, eds, *Modern Times: Reflections on a Century of English Modernity* (London, Routledge, 1996).

Nevett, T. R., *Advertising in Britain: A History* (London, Heinemann, 1982).

Nicholls, Peter, 'Futurism, gender, and theories of postmodernity', *Textual Practice* 3: 2 (summer 1989), 210–11.

Nicholls, Peter, *Modernisms: A Literary Guide* (Basingstoke, Macmillan, 1995).

Nolan, Emer, *James Joyce and Nationalism* (London and New York, Routledge, 1995).

Norquay, Glenda and Gerry Smyth, eds, *Space and Place: The Geographies of Literature* (Liverpool, Liverpool John Moores University Press, 1997).

North, Michael, *Reading 1922: A Return to the Scene of the Modern* (Oxford, Oxford University Press, 1999).

O'Connell, Sean, *The Car and British Society: Class, Gender and Motoring, 1896–1939* (Manchester, Manchester University Press, 1998).

O'Connor, Teresa, 'Jean Rhys, Paul Theroux, and the Imperial Road', *Twentieth Century Literature* 38: 4 (winter 1992), 404–14.

O'Connor, Teresa, *Jean Rhys: The West Indian Novels* (New York and London, New York University Press, 1986).

O'Toole, Fintan, 'Going west: the country versus the city in Irish writing', *The Crane Bag* 9: 2 (1985), 111–16.

Osborne, Peter, 'Modernity is a qualitative, not a chronological, category', *New Left Review* 192 (March–April 1992), 65–84.

Ossory Fitzpatrick, Samuel A., *Dublin: A Historical and Topographical Account of the City* (1907) (Dublin: Tower Books, 1977).

Overy, Richard, 'Heralds of modernity: cars and planes from invention to necessity', in M. Teich and Roy Porter, eds, Fin de Siècle *and its Legacy* (Cambridge, Cambridge University Press, 1990).

Parsons, Deborah L., *Streetwalking the Metropolis: Women, the City and Modernity* (Oxford, Oxford University Press, 2000).

Paulin, Tom, 'A new look at the language question', in Field Day Theatre Company, eds, *Ireland's Field Day* (London, Hutchinson, 1985).

Pennell, Joseph, 'Motors and cycles: the transition stage', *The Contemporary Review* (1 February 1902), 185–95.

Peters Corbett, David and Andrew Thacker, 'Cultural formations in modernism: movements and magazines 1890–1920', *Prose Studies* 16: 2 (August 1993), 84–106.

Pevsner, Nikolas, *The Buildings of England: London I – The Cities of London and Westminster* (Harmondsworth, Penguin, 1957).

Philo, C., 'Foucault's geography', *Envirnoment and Planning D: Society and Space* 10 (1992), 137–61.

Pile, Steve, *The Body and the City: Psychoanalysis, Space and Subjectivity* (London and New York, Routledge, 1996).

Pinkney, Tony, 'Space: the final frontier', *News from Nowhere* 8 (1990), 10–27.

Plowden, William, *The Motor Car and Politics in Britain* (Harmondsworth, Penguin, 1973).

Pollock, Griselda, *Vision and Difference: Feminity, Feminism and Histories of Art* (London and New York, Routledge, 1988).
Pondrom, Cyrena N., 'H. D. and the origins of Imagism', in Susan Stanford Friedman and Rachel Blau DuPleissis, eds, *Signets: Reading H. D.* (Wisconsin, University of Wisconsin Press, 1990).
Porter, Bernard, *The Lion's Share: A Short History of British Imperialism 1850–1983*, 2nd edn (London, Longman, 1984)
Pound, Ezra, 'The new sculpture', *The Egoist* (February 1914), 68.
Pound, Ezra, *ABC of Reading* (London, Routledge & Kegan Paul, 1934).
Pound, Ezra, *Gaudier-Brzeska: A Memoir* (1916) (New York, New Directions, 1970).
Pound, Ezra, *Selected Prose 1909–1965*, ed. William Cookson (London, Faber, 1973).
Pound, Ezra, *Collected Shorter Poems* (London, Faber, 1984).
Pykett, Lyn, *Engendering Fictions: The English Novel in the Early Twentieth Century* (London, Edward Arnold, 1995).
Raban, Jonathan, *Soft City* (London, Collins Harvill, 1988).
Rabkin, Eric S., 'Spatial form and plot', *Critical Inquiry* 4 (1977–78), 253–70.
Rainey, Lawrence, *Institutions of Modernism: Literary Elites and Public Culture* (New Haven, CT, and London, Yale University Press, 1998).
Raitt, Suzanne, *Vita and Virginia: The Work and Friendship of Vita Sackville-West and Virginia Woolf* (Oxford, Clarendon Press, 1993).
Randall, Bryony, 'John Gould Fletcher's city aesthetic: "London Excursion"', unpublished paper (n.d.).
Rhys, Jean, *After Leaving Mr Mackenzie*, British Library, BL Add. MS 57856.
Rhys, Jean, *The Left Bank and Other Stories* (London, Jonathan Cape, 1927).
Rhys, Jean, *Wide Sargasso Sea* (Harmondsworth, Penguin, 1968).
Rhys, Jean, *Good Morning, Midnight* (Harmondsworth, Penguin, 1969).
Rhys, Jean, 'Temps Perdi', *Penguin Modern Stories 1*, ed. Judith Burnley (Harmondsworth, Penguin, 1969).
Rhys, Jean, *Voyage in the Dark* (Harmondsworth, Penguin, 1969).
Rhys, Jean, *After Leaving Mr Mackenzie* (Harmondsworth, Penguin, 1971).
Rhys, Jean, *Tigers Are Better Looking, with a Selection from The Left Bank* (Harmondsworth, Penguin, 1972).
Rhys, Jean, *Quartet* (Harmondsworth, Penguin, 1973).
Rhys, Jean, *Sleep it Off, Lady* (London, Andre Deutsch, 1976).
Rhys, Jean, *Smile, Please: An Unfinished Autobiography* (Harmondsworth, Penguin, 1981).
Rhys, Jean, *Letters 1931–1966*, ed. Francis Wyndham and Diana Melly (Harmondsworth, Penguin, 1985).
Rhys, Jean, '*Voyage in the Dark* part IV (original version)', in Bonnie Kime Scott, ed., *The Gender of Modernism: A Critical Anthology* (Bloomington, Indiana University Press, 1990).

Richards, I. A., *Practical Criticism: A Study of Literary Judgement* (1929) (London, Routledge, 1964).

Richards, Jeffrey and John M. MacKenzie, *The Railway Station: A Social History* (Oxford, Oxford University Press, 1986).

Richardson, Dorothy, *Pilgrimage* (London, Virago, 1979), vol.1.

Riquelme, John Paul, *Teller and Tale in Joyce's Fiction* (Baltimore, Johns Hopkins University Press, 1983).

Roberts, Peter, *The Motoring Edwardians* (London, Ian Allan, 1978).

Robinson, Alan, *Poetry, Painting and Ideas 1885–1914* (London, Macmillan, 1985).

Roe, Sue, '"The shadow of light": the symbolic underworld of Jean Rhys', in Sue Roe, ed., *Women Reading Women's Writing* (Brighton, Harvester Press, 1987).

Ross, Kristin, *The Emergence of Social Space: Rimbaud and the Paris Commune* (London, Macmillan, 1988).

Ruskin, John, *The Seven Lamps of Architecture* (London, George Allen, 1907).

Ruthven, K. K., *A Guide to Ezra Pound's 'Personae' (1926)* (Berkeley and Los Angeles, University of California Press, 1969).

Said, Edward, *Culture and Imperialism* (London, Vintage, 1994).

Saler, Michael T., *The Avant-Garde in Interwar England: Medieval Modernism and the London Underground* (Oxford, Oxford University Press, 1999).

Salusinszky, Imre, ed., *Criticism in Society* (New York and London, Methuen, 1987)

Saunders, Ann, *The Art and Architecture of London: An Illustrated Guide* (Oxford, Phaidon, 1984).

Savory, Elaine, *Jean Rhys*, Cambridge Studies in African and Caribbean Literature (Cambridge, Cambridge University Press, 1998).

Schivelbusch, Wolfgang, *The Railway Journey: The Industrialization of Time and Space in the Nineteenth Century* (Leamington Spa and New York, Berg, 1986).

Schnapp, Jeffrey T., 'Crash (speed as engine of individuation)', *Modernism/Modernity* 6: 1 (January 1999), 1–50.

Schwartz, Sanford, *The Matrix of Modernism: Pound, Eliot, and Early Twentieth Century Thought* (Princeton, NJ, Princeton University Press, 1985).

Schwarz, Daniel, *The Transformation of the English Novel 1890–1930* (Basingstoke, Macmillan, 1989),

Scott, Bonnie Kime, 'James Joyce: a subversive geography of gender', in Paul Hyland and Neil Sammells, eds, *Irish Writing: Exile and Subversion* (Basingstoke, Macmillan, 1991).

Scott, Bonnie Kime, *Refiguring Modernism*, 2 vols (Bloomington, Indiana University Press, 1995).

Seamus Heaney, 'A sense of place' in *Preoccupations: Selected Prose 1969–1978* (London, Faber, 1980).

Seidel, Michael, *Epic Geography: James Joyce's 'Ulysses'* (Princeton, NJ, Princeton University Press, 1976).

Selcon, G. A., 'Pullman cars on the Metropolitan Railway', *The Railway and Travel Monthly*, 1: 3 (July 1910), 228–35.

Sennett, Richard, *The Conscience of the Eye: The Design and Social Life of Cities* (London, Faber, 1990).

Sennett, Richard, *Flesh and Stone: The Body and the City in Western Civilization* (London, Faber, 1994).

Shields, Rob, *Places on the Margin: Alternative Geographies of Modernity* (London, Routledge, 1991).

Showalter, Elaine, *The Female Malady: Women, Madness and English Culture, 1830–1980* (London, Virago, 1987).

Simmel, Georg, 'The metropolis and mental life' in *The Sociology of Georg Simmel*, trans. and ed. Kurt H Wolff (New York, Free Press, 1950).

Simmons, Jack, *The Victorian Railway* (London, Thames & Hudson, 1991).

Smyth, Gerry, *The Novel and the Nation: Studies in the New Irish Fiction* (London, Pluto, 1997).

Soja, Edward W., *Postmodern Geographies: The Reassertion of Space in Critical Social Theory* (London and New York, Verso, 1989).

Soja, Edward W., *Thirdspace: Journeys to Los Angeles and Other Real-and-Imagined Places* (Oxford, Blackwell, 1996).

Some Imagist Poets: An Anthology (Boston and New York, Houghton Mifflin, 1915).

Some Imagist Poets 1916: An Annual Anthology (Boston and New York, Houghton Mifflin, 1916).

Some Imagist Poets 1917: An Annual Anthology (Boston and New York, Houghton Mifflin, 1917).

Squier, Susan M., *Virginia Woolf and London: The Sexual Politics of the City* (Chapel Hill and London, University of North Carolina Press, 1985).

Squier, Susan M., 'Virginia Woolf's London and the feminist revision of modernism', in Mary Ann Caws, ed., *City Images: Perspectives from Literature, Philosophy, and Film* (New York, Gordon & Breach, 1991).

Stansky, Peter, *On or About December 1910: Early Bloomsbury and its Intimate World* (Cambridge, MA, Harvard University Press, 1997).

Stead, C. K., *The New Poetic: Yeats to Eliot* (Harmondsworth, Penguin, 1967).

Stevenson, John, *British Social History 1914–45* (Harmondsworth, Penguin, 1984).

Stevenson, Randall, *Modernist Fiction: An Introduction* (Hemel Hempstead, Harvester Wheatsheaf, 1992).

Stewart, Jack F., 'Spatial form and color in *The Waves*', *Twentieth Century Literature* 28: 1 (spring 1982), 86–107.

Stock, Noel, *The Life of Ezra Pound* (Harmondsworth, Penguin Books, 1974).

Stone, Wilfred, *The Cave and the Mountain: A Study of E. M. Forster* (London, Oxford University Press, 1966).

Sutherland, John, *Can Jane Eyre Be Happy?* (Oxford, Oxford University Press, 1997).

Tambling, Jeremy, ed., *E. M. Forster: New Casebook* (Basingstoke, Macmillan, 1995).

Tate, Trudi, *Modernism, History and the First World War* (Manchester, Manchester University Press, 1998).

Tester, Keith, ed., *The Flâneur* (London and New York, Routledge, 1994).

Teyssot, Georges, 'Heterotopias and the history of spaces', *Architecture and Urbanism* 121 (1980), 79–100.

Thacker, Andrew, 'Amy Lowell and H. D.: the other Imagists', *Women: A Cultural Review* 4: 1 (spring 1993), 49–59.

Thacker, Andrew, 'Dora Marsden and *The Egoist*: "our war is with words"', *English Literature in Transition 1880–1920* 36: 2 (1993), 179–96.

Thacker, Andrew, 'Journey with maps: travel theory, geography and the syntax of space', in Charles Burdett and Derek Duncan, eds, *Cultural Encounters: European Travel Writing in the 1930s* (Oxford, Berghahn, forthcoming).

Thacker, Andrew, 'Unrelated beauty: polyphonic prose and the Imagist city', in Adrienne Munich and Melissa Bradshaw, eds, *Presenting Amy Lowell: Critical Essays* (New Jersey, Rutgers University Press, forthcoming).

Thomas, J. P., *Handling London's Underground Traffic* (London, London's Underground, 1928).

Tickner, Lisa, *The Spectacle of Women: Imagery of the Suffrage Campaign 1907–14* (London, Chatto & Windus, 1987).

Timms, E. and D. Kelly, eds, *Unreal City: Urban Experience in Modern European Literature and Art* (Manchester, Manchester University Press, 1985).

Torchiana, Donald T., *Backgrounds for Joyce's 'Dubliners'* (Boston, Allen & Unwin, 1986).

Trilling, Lionel, *E. M. Forster: A Study* (London, Hogarth, 1969).

Trotter, David, *The English Novel in History 1895–1920* (London, Routledge, 1993).

Urry, John, *The Tourist Gaze: Leisure and Travel in Contemporary Societies* (London, Sage, 1990).

Virilio, Paul, *The Art of the Motor*, trans. Julie Rose (Minneapolis and London, University of Minnesota Press, 1995.

Watson, Sophie and Katherine Gibson, eds, *Postmodern Cities and Spaces* (Oxford, Blackwell, 1995).

Wharton, Edith, *A Motor Flight Through France* (1908) (London, Picador, 1995).

Wheeler, Kathleen, *'Modernist' Women Writers and Narrative Art* (Basingstoke, Macmillan, 1994).

Widdowson, Peter, *E. M. Forster's 'Howards End': Fiction as History* (Brighton, Sussex University Press, 1977).

Williams, Raymond, *The Country and the City* (Oxford/New York, Oxford University Press, 1973).
Williams, Raymond, 'The Bloomsbury fraction' in *Problems in Materialism and Culture* (London, Verso, 1980).
Williams, Raymond, *The Politics of Modernism: Against the New Conformists* (London: Verso, 1989).
Williams, Rosalind, *Notes on the Underground: An Essay on Technology, Society, and the Imagination* (Cambridge, MA, MIT Press, 1990).
Wilson, Elizabeth, *The Sphinx in the City: Urban Life, the Control of Disorder and Women* (London, Virago, 1991).
Wilson, Elizabeth, 'The invisible *flâneur*', *New Left Review* 191 (January–February 1992), 90–110.
Wilson, Elizabeth, 'The rhetoric of urban space', *New Left Review* 209 (January–February 1995), 146–60.
Wolff, Janet, 'The invisible *flâneuse*: women and the literature of modernity', in Andrew Benjamin, ed., *The Problems of Modernity: Adorno and Benjamin* (London, Routledge, 1989).
Wolff, Janet, 'On the road again: metaphors of travel in cultural criticism' in *Resident Alien: Feminist Cultural Criticism* (Cambridge, Polity Press, 1995).
Woodruff, William, *The Rise of the British Rubber Industry during the Nineteenth Century* (Liverpool, Liverpool University Press, 1958).
Woolf, Leonard, *Downhill All the Way: An Autobiography of the Years 1919–1939* (London, Hogarth Press, 1967).
Woolf, Virginia, *A Haunted House and Other Stories* (Harmondsworth, Penguin, 1973).
Woolf, Virginia, *Mrs Dalloway's Party: A Short Story Sequence*, ed. Stella McNichol (London, Hogarth Press, 1973).
Woolf, Virginia, *The Letters of Virginia Woolf*, 6 vols, ed. Nigel Nicholson and Joanne Trautmann (London, Hogarth Press, 1975–80).
Woolf, Virginia, *Moments of Being: Unpublished Autobiographical Writings*, ed. Jeanne Schulkind (Sussex, University Press, 1976).
Woolf, Virginia, *Mrs Dalloway* (London, Granada, 1976).
Woolf, Virginia, *A Room of One's Own* (London, Granada, 1977).
Woolf, Virginia, *The Waves* (London, Granada, 1977).
Woolf, Virginia, *To the Lighthouse* (London, Granada, 1977).
Woolf, Virginia, *The Diary of Virginia Woolf*, 5 vols, ed. Anne Oliver Bell and Andrew McNeillie (London, Penguin, 1982).
Woolf, Virginia, *Virginia Woolf's Reading Notebooks*, ed. Brenda R. Silver (Princeton, NJ, Princeton University Press, 1983).
Woolf, Virginia, *The Essays of Virginia Woolf*, 6 vols, ed. Andrew McNeillie (London, Hogarth Press, 1986).
Woolf, Virginia, *Jacob's Room*, ed. Sue Roe (London, Penguin, 1992).
Woolf, Virginia, *Orlando: A Biography* (London, Vintage, 1992).

Woolf, Virginia, *A Woman's Essays: Selected Essays, volume 1*, ed. Rachel Bowlby (London, Penguin, 1992).
Woolf, Virginia, *The Crowded Dance of Modern Life: Selected Essays, volume 2*, ed. Rachel Bowlby (London, Penguin, 1993).
Woolf, Virginia, *Travels with Virginia Woolf*, ed. Jan Morris (London, Pimlico, 1997).
Wright, Anne, *Literature of Crisis 1910–22* (London, Macmillan, 1984).
Yeats, W. B., *Yeats's Poems*, ed. A. Norman Jeffares (London, Macmillan, 1989).
Zemgulys, Andrea P., '"*Night and Day* is dead": Virginia Woolf in London "literary and historic"', *Twentieth-Century Literature* 46: 1 (spring 2000), 56–77.

Index

Note: illustrations are represented by italics; 'n.' after a page reference indicates a note on that page; all literary works are listed under name of author.

Lightning Source UK Ltd.
Milton Keynes UK
UKHW04f1902030918
328271UK00001B/43/P

9 780719 081200